MAMMALIAN
CELL GENETICS

CELL BIOLOGY: A SERIES OF MONOGRAPHS

E. Edward Bittar, Series Editor

MAMMALIAN CELL GENETICS

MARTIN L. HOOPER
Department of Pathology
University Medical School
Edinburgh, United Kingdom

A WILEY-INTERSCIENCE PUBLICATION

JOHN WILEY & SONS

New York • Chichester • Brisbane • Toronto • Singapore

Library of Congress Cataloging in Publication Data:

Hooper, Martin L.
  Mammalian cell genetics.

  (Cell biology ; v. 4)
  "A Wiley-Interscience publication."
  Includes bibliographical references and index.
    1. Mammals—Genetics. 2. Cytogenetics. I. Title.
II. Series: Cell biology (New York, N.Y.) ; v. 4.

QL738.5.H66   1985   599'.0873223   84-17425
ISBN 0-471-89201-7

Printed in the United States of America

10 9 8 7 6 5 4 3 2 1

SERIES PREFACE

The aim of the Cell Biology Series is to focus attention upon basic problems and show that cell biology as a discipline is gradually maturing. In its largest aim, each monograph seeks to be readable and informative, scholarly, and the work of a single mind. In general, the topics chosen deal with major contemporary issues. Together they represent a rather large domain whose importance has grown enormously in the course of the last generation. The introduction of new techniques has no doubt ushered in a small revolution in cell biology. However, we still know very little about the cell as an ordered structure. As will become abundantly clear to the reader, real progress is not just a matter of progress of technique but also a matter of close interaction between advances in different fields of study, as well as genesis of new approaches and generalized concepts.

E. EDWARD BITTAR

Madison, Wisconsin
January 1984

PREFACE

This book is concerned with the genetic analysis of cells taken from a multicellular mammal and cultured as though they were independent microorganisms. This form of analysis has come to be known as somatic cell genetics, although its application is not restricted to somatic cells. My aim was not to write a textbook. The student requiring an introduction to somatic cell genetics is already well served by the books of Puck (1) and Littlefield (2) which describe the early development of the subject, and by the recent, more comprehensive texts of Morrow (3) and Martin et al. (4). Rather, I have tried to bridge the gap between such works and multiauthor volumes such as those edited by Caskey and Robbins (5) and Shay (6) which describe selected aspects of the subject in depth. This book is aimed at research workers who either are about to enter the field of mammalian somatic cell genetics, or have experience in one area of the field but require a guide to the literature in other parts of it. To keep the size of the book within manageable bounds, I have set myself the limited aim of making the reader aware of work described elsewhere rather than of explaining it in detail. In other words, the book is designed not to stand alone but to be read in conjunction with the primary literature. This explains the lack of explanatory diagrams, which would be a serious omission in a more introductory text, and the relative profusion of tables.

Somatic cell genetics has close relationships both with microbial genetics, from which it largely derives its strategies and approaches, and with the classical genetics of higher eukaryotes, with which it shares its aims. For the genetic analysis of mammals the somatic cell genetic approach has certain advantages compared to the classical approach: The environment of the individual cell is better controlled in culture, and the generation time of the cell is much shorter than that of the whole organism. It also has drawbacks: First, the very fact that the culture environment differs from a cell's environment in the whole organism imposes selection in favor of abnormal cells; second, cells *in vivo* interact with other cells of the same organism both through direct cell contact and via diffusible signal molecules. Some such interactions occur in cell culture, limiting the extent to which cells behave as independent microorganisms. Other interactions do not occur in cell culture, limiting the range of phenomena that can be studied. Nevertheless, because the techniques available to the somatic cell geneticist and the classical geneticist are different, the kinds of information available from the two approaches are often complementary, and systems where they can be combined are particularly powerful (see Chapter 6).

The text is organized in terms of techniques and approaches rather than in terms of applications. Access to a particular application may be made from the Index. Chapter 1 gives an overview of the basic techniques of mammalian cell culture, although some knowledge in this area is assumed and emphasis is placed upon those aspects of direct relevance to later chapters. Chapter 2 discusses selective techniques, including some which have not yet been extensively applied in somatic cell genetics but have potential for future work. Chapters 3 and 4 deal with the central topics of cell variants and cell hybrids. Chapter 5 is concerned with a major growth area, namely, the application of the techniques of molecular genetics to cultured cells. The inclusion of Chapter 6 is partly a reflection of my own interest in teratocarcinomas but mainly a consequence of the potential of work in this area for unifying the genetics of the cell with that of the organism. I have sought to emphasize the relationships between the different branches of the subject, with the more classical work described in the first four chapters providing a link between the molecule and the multicellular organism.

For the convenience of the reader I have not given references to the primary literature in all cases but have often cited reviews from which the original references may obtained. I hope that authors who, for this reason, have not been acknowledged will forgive me.

There is as yet no universally accepted system of abbrevations in use in somatic cell genetics. The one I have adopted is consistent with the majority of current usage: Gene products are given capital letters, for example, TK (thymidine kinase); variant phenotypes are also given in capitals, for example, TK^- (thymidine kinase defective) and OUA^R (ouabain-resistant); gene loci are given in lower case italic letters, for example, *tk*; complementation groups are given with an initial capital letter, for example, UrdA.

I am grateful to many colleagues for valuable discussions that have shaped this book. The idea of writing it can be traced back to a conversation with Ben Carritt some years ago. Professor Sir Alastair Currie and Austin Smith kindly read the manuscript and made a number of constructive suggestions. Kit Gardner provided expert secretarial assistance. The shortcomings, which may remain, are entirely my own responsibility and I hope that readers will be kind enough to point them out to me.

MARTIN L. HOOPER

Edinburgh, United Kingdom
October 1984

CONTENTS

MAMMALIAN CELL GENETICS

1

THE CULTURED MAMMALIAN CELL

The methods of mammalian cell genetics have been developed by combining strategies largely derived from microbial genetics with the techniques of mammalian cell culture. It is assumed that the reader is familiar with basic mammalian cell culture techniques, therefore only a broad overview is given here to put into context those aspects of particular relevance to later chapters. More detailed information will be found in references 7–11. The composition of growth media (Section 1.1) forms the basis of the majority of the selective systems described in Chapter 2. The requirement of most cell types for attachment to a substratum (Section 1.2) also constrains the design of selective systems, particularly in selecting for cells present at very low frequencies in a population where a large input of cells, and therefore a large surface area, is required. Subculturing (Section 1.4), cloning (Section 1.9), and cryopreservation (Section 1.10) of selected cells are also of paramount technical importance. The remainder of the chapter deals with the relationship between tissue-culture cells and cells in normal tissues. Most of the techniques of mammalian cell genetics cannot be carried out directly on primary cultures (Section 1.3) but require the establishment of permanent lines (Section 1.5). The relationship of cells of such lines to normal cells in chromosome constitution, state of differentiation, and interaction with other cells is discussed in Sections 1.6, 1.7, and 1.8. To date, the majority of work in somatic cell genetics has been with "fibroblast-like" cells (Section 1.5), but cell lines are now available that exhibit different states of differentiation and are suitable for use in somatic cell genetics (Section 1.7).

1.1. GROWTH MEDIA

Early tissue-culture experiments used biological fluids such as serum as growth media. As biochemical characterization of serum components proceeded, numerous formulations of chemically defined media were developed for particular cell types (reviewed by Waymouth, ref. 12). Although some of these were designed to reduce the requirement for serum protein to minimal levels none is adequate for completely serum-free growth of any cell line with the exception of certain lines that have undergone strong selection for ability to grow *in vitro* and are as a consequence highly aneuploid (Sections 1.5 and 1.6). For this reason relatively simple media have become popular, such as Eagle's minimum essential medium (MEM, Table 1.1) which was designed to reduce the requirement for amino acids and vitamins to the minimum necessary for use with a supplement of 5–10% serum. Nonetheless, considerable effort has been devoted to defining all the low-molecular-weight growth-promoting molecules contributed by serum and this has led to the development of media such as Ham's F12 and more recently MDCB105 (Table 1.1) which are sufficiently complete to allow the serum requirement to be fulfilled for certain cell types by purified hormones or macromolecular serum fractions (Section 1.1.9). The closer one approaches a completely defined medium, the more marked become the differences in the requirement of different cell types, which emphasizes the point that the growth of one cell type in medium whose composition has been optimized for another, which is a widespread practice in tissue-culture laboratories, may impose a considerable selective pressure in favor of abnormal cells.

The majority of normal cell types show anchorage-dependent growth, although cell lines capable of growth in suspension can be obtained from normal and malignant hematopoietic cells (14) and by adaptation of monolayer cell lines such as HeLa and L cells (7). Media for suspension cultures are very similar to those used for monolayer culture, the principal modifications being a lowering of the Ca^{2+} ion concentration, the use of somewhat higher concentrations of nutrients and serum, and the incorporation of carboxymethyl cellulose to overcome precipitation problems at the gas/liquid interface (7,15).

1.1.1. Carbon Source

The most commonly used carbon source in cell culture media is glucose, although many other sugars can be used instead, and galactose has the advantage that its metabolites lead to less acidification of the medium than do those of glucose (12,19). In addition, sodium pyruvate is commonly added to media designed for use at low cell density as unlike the phosphorylated intermediates of glycolysis it is readily lost from the cytoplasm to the external medium if the concentration gradient across the cell membrane is high.

1.1.2. Major Salts, Buffers, and the Gas Phase

The principal constraints determining the concentrations of major inorganic ions in tissue-culture media are total osmolarity and ionic balance (19).

Media are generally designed to be isoosmotic with serum, although deviations of up to 10% from the nominal osmolarity appear to be well tolerated, at least by L cells. Na/K ratios appear to be particularly important, and different ratios are optimal for different cell lines. For pH maintenance bicarbonate buffers requiring equilibration with an atmosphere of 5–10% CO_2 are most commonly employed. Use of other buffers is complicated by the fact that cells appear to have a metabolic requirement for CO_2 in addition to a requirement for pH maintenance, and also that tris and phosphate buffers used alone exhibit some toxicity at concentrations required to ensure adequate buffering. In order to dispense with the need for gassing with CO_2/air, Leibovitz (20) employed phosphate and high concentrations of arginine and histidine to achieve buffering, together with pyruvate to generate endogenous CO_2 and galactose as a carbon source to reduce the production of lactic acid. Other successful strategies include the use of β-glycerophosphate or HEPES to provide major buffering capacity with the addition of a low concentration of bicarbonate (19,21,22).

Oxygen tension is also an important variable (19,22). Although 5% CO_2/ 95% air is most commonly used for gassing cell cultures there are indications that the optimal oxygen concentration in the gas phase may be as low as 1–3% (23), and hyperbaric oxygen levels such as are helpful for maintenance of tissue fragments are markedly toxic to cell cultures (19,22). This toxicity is most severe under conditions of selenium deficiency (see Section 1.1.6). Vitamin E may exert a protective effect (24).

1.1.3. Amino Acids

In general ten amino acids are essential in the mammalian diet (Table 1.1) and these are included in tissue-culture media together with cysteine (or cystine), glutamine, and tyrosine, whose synthesis is restricted to certain tissues only and which are therefore essential for the majority of cell types. In addition media designed for use at clonal cell density include the remaining seven amino acids of the twenty which are protein constituents. These can be synthesized by cells but are readily lost to the medium if the transmembrane concentration gradient is high (compare pyruvate, Section 1.1.1).

1.1.4. Vitamins

Whereas the B vitamins (Table 1.1) are essential for the growth of cultured cells, the fat-soluble vitamins appear not to be (24). Ascorbate may have

TABLE 1.1. COMPOSITION OF DEFINED MEDIA FOR CULTURE OF MAMMALIAN CELLS

Medium Components	MEM (16) (moles/L)	F12 (17) (moles/L)	MDCB105 (18) (moles/L)
Carbon Sources			
D-glucose	5.5×10^{-3}	1.0×10^{-2}	4.0×10^{-3}
Pyruvate (Na salt)		1.0×10^{-3}	1.0×10^{-3}
Major Salts, Buffers, and Indicators			
$CaCl_2 \cdot 2H_2O$	1.8×10^{-3}	3.0×10^{-4}	1.0×10^{-3}
KCl	5.4×10^{-3}	3.0×10^{-3}	
KH_2PO_4			3.0×10^{-3}
$MgCl_2 \cdot 6H_2O$	1.0×10^{-3}	6.0×10^{-4}	
$MgSO_4 \cdot 7H_2O$			1.0×10^{-3}
NaCl	1.16×10^{-1}	1.3×10^{-1}	1.12×10^{-1}
$NaH_2PO_4 \cdot 2H_2O$	1.1×10^{-3}		
$Na_2HPO_4 \cdot 7H_2O$		1.0×10^{-3}	3.0×10^{-3}
HEPES			3.0×10^{-2}
$NaHCO_3$	2.38×10^{-2}	1.4×10^{-2}	
NaOH			2.0×10^{-2}
Phenol red	4.0×10^{-5}	3.3×10^{-6}	3.3×10^{-6}
Amino Acids			
(i) Essential in Mammalian Diet			
L-Arginine·HCl	6.0×10^{-4}	1.0×10^{-3}	1.0×10^{-3}
L-Histidine·HCl·H_2O	2.0×10^{-4}	1.0×10^{-4}	1.0×10^{-4}
L-Isoleucine	4.0×10^{-4}	3.0×10^{-5}	3.0×10^{-5}
L-Leucine	4.0×10^{-4}	1.0×10^{-4}	1.0×10^{-4}
L-Lysine·HCl	4.0×10^{-4}	2.0×10^{-4}	2.0×10^{-4}
L-Methionine	1.0×10^{-4}	3.0×10^{-5}	3.0×10^{-5}
L-Phenylalanine	2.0×10^{-4}	3.0×10^{-5}	3.0×10^{-5}
L-Threonine	4.0×10^{-4}	1.0×10^{-4}	1.0×10^{-4}
L-Tryptophan	5.0×10^{-5}	1.0×10^{-5}	1.0×10^{-5}
L-Valine	4.0×10^{-4}	1.0×10^{-4}	1.0×10^{-4}
(ii) Essential for Most Cell Types			
L-Cysteine·HCl·H_2O	2.0×10^{-4}	2.0×10^{-4}	5.0×10^{-5}
L-Glutamine	2.0×10^{-3}	1.0×10^{-3}	2.5×10^{-3}
L-Tyrosine	2.0×10^{-4}	3.0×10^{-5}	3.0×10^{-5}

(*Table continues on p. 6.*)

Table 1.1 (*continued*)

Medium Components	MEM (16) (moles/L)	F12 (17) (moles/L)	MDCB105 (18) (moles/L)
(iii) Nonessential			
L-Alanine		1.0×10^{-4}	1.0×10^{-4}
L-Asparagine·H_2O		1.0×10^{-4}	1.0×10^{-4}
L-Aspartic acid		1.0×10^{-4}	1.0×10^{-4}
L-Glutamic acid		1.0×10^{-4}	1.0×10^{-4}
Glycine		1.0×10^{-4}	1.0×10^{-4}
L-Proline		3.0×10^{-4}	3.0×10^{-4}
L-Serine		1.0×10^{-4}	1.0×10^{-4}
Vitamins			
d-Biotin		3.0×10^{-8}	3.0×10^{-8}
Folic acid	2.3×10^{-6}	3.0×10^{-6}	
Folinic acid (Ca leucovorin)			1.0×10^{-9}
DL-α-Lipoic acid		1.0×10^{-6}	1.0×10^{-8}
Niacinamide	8.2×10^{-6}	3.0×10^{-7}	5.0×10^{-5}
D-Pantothenic acid (hemi Ca salt)	4.6×10^{-6}	1.0×10^{-6}	1.0×10^{-6}
Pyridoxal	6.0×10^{-6}		
Pyridoxine·HCl		3.0×10^{-7}	3.0×10^{-7}
Riboflavin	2.7×10^{-7}	1.0×10^{-7}	1.0×10^{-7}
Thiamine	3.0×10^{-6}	1.0×10^{-6}	1.0×10^{-6}
Vitamin B_{12}		1.0×10^{-6}	1.0×10^{-7}
Other Organic Compounds			
Adenine			1.0×10^{-5}
Choline chloride	8.2×10^{-6}	1.0×10^{-4}	1.0×10^{-4}
Hypoxanthine		3.0×10^{-5}	
i-Inositol	1.1×10^{-5}	1.0×10^{-4}	1.0×10^{-4}
Linoleic acid		3.0×10^{-7}	1.0×10^{-8}
Putrescine·2HCl		1.0×10^{-6}	1.0×10^{-9}
Thymidine		3.0×10^{-6}	3.0×10^{-7}

Table 1.1 (*continued*)

Medium Components	MEM (16) (moles/L)	F12 (17) (moles/L)	MDCB105 (18) (moles/L)
		Trace Elements	
$CuSO_4 \cdot 5H_2O$		1.0×10^{-8}	1.0×10^{-9}
$FeSO_4 \cdot 7H_2O$		3.0×10^{-6}	5.0×10^{-6}
H_2SeO_3			3.0×10^{-8}
$MnSO_4 \cdot 5H_2O$			1.0×10^{-9}
$Na_2SiO_3 \cdot 9H_2O$			5.0×10^{-7}
$(NH_4)_6Mo_7O_{24} \cdot 4H_2O$			1.0×10^{-9}
NH_4VO_3			5.0×10^{-9}
$NiCl_2 \cdot 6H_2O$			5.0×10^{-10}
$SnCl_2 \cdot 2H_2O$			5.0×10^{-10}
$ZnSO_4 \cdot 7H_2O$		3.0×10^{-6}	5.0×10^{-7}

a beneficial effect, especially for collagen-synthesizing cells, but its instability to oxidation has led to conflicting reports about its effectiveness (15). In media such as MEM supplemented with 5–10% serum, adequate levels of some of the B vitamins are provided by the serum.

1.1.5. Other Organic Compounds

Choline and i-inositol are required by most cultured cells. Purine and pyrimidine sources may be beneficial where folate is in short supply or inefficiently used. A source of polyunsaturated fatty acid is required by most cell types: In serum-containing media this is normally satisfied by lipid bound to serum protein but for growth in serum-free media or media containing delipidized serum (see below) addition of linoleic acid is necessary. In addition some cell types show a requirement for a monounsaturated fatty acid (25). Putrescine or other polyamines are required for clonal growth of Chinese hamster ovary cells in protein-free media (26).

Some hematopoietic tumor cells have been found to require a thiol compound for growth in serum-containing medium (27) and addition of a thiol such as β-mercaptoethanol has also been found to be beneficial for the growth of embryonal carcinoma cells (28). Thiol-requiring hematopoietic tumor cells are defective in the enzyme methylthioadenosine nucleoside phosphorylase (MTANP) which produces methylthioribose 1-phosphate from methylthioadenosine. Methylthioribose 1-phosphate is degraded via 2-keto 4-methyl thiobutyric acid to methyl mercaptan. Fetal calf serum contains methyl mercaptan bound to protein by disulfide linkage, and growth of thiol-requiring cells can be obtained in serum-free medium using either a combination of a reactive methylthio group donor molecule and a low concentration of a thiol, or a mixed disulfide of methyl mercaptan with cysteine, thioglycerol, glutathione, or thioglycolic acid in the absence of a free thiol compound. Therefore, it appears that cells require methyl mercaptan which is generated endogenously in MTANP$^+$ cells but must be supplied exogenously in MTANP$^-$ cells, and that thiol compounds function by liberating methyl mercaptan from disulfide linkage to serum protein and then serving as carriers to take it through the cell membrane as mixed disulfides (29–32, and unpublished results cited therein).

1.1.6. Trace Elements

With the use of purer reagents for media preparation, progress toward defined serum-free media has revealed requirements for a number of trace elements (33,34; Table 1.1). The effects of some of these are small and variable: For example, it is difficult to demonstrate an effect of added nickel salts because nickel is a common contaminant in most reagents. However, the requirements for iron and selenium are readily demonstrated and are of particular interest in forming the basis of certain selective media (Section 2.1.5). Selenium is an integral part of the enzyme glutathione peroxidase, which destroys lipid peroxides (35). This explains its role in protecting against oxygen toxicity (Section 1.1.2) and its nutritional interrelationship with vitamin E, which also functions as an antioxidant (24,33).

1.1.7. Antibiotics

Penicillin and streptomycin are commonly added to tissue-culture media to discourage the growth of microorganisms. Their use, however, may be counterproductive as they may mask the overt symptoms of microbial infection while permitting less readily detectable contaminants such as mycoplasma (36) to become established. Mycoplasma are insensitive to penicillin because they lack a cell wall and can cause a wide range of artifacts in infected cultures. In addition some toxic effects of antibiotics have been noted in serum-depleted media (12). Antifungal antibiotics such as mycostatin and amphotericin B (fungizone) are not added routinely to media but are retained for use if a contamination arises in a valuable stock, as concentrations which are effective against contaminants are also somewhat toxic to cells (7). As a rule antibiotics are best avoided completely: Good sterile technique can reduce the incidence of contamination to a level where one can discard infected cultures and return to clean frozen stocks.

1.1.8. Serum

The problems experienced in designing media for growth of cells in the complete absence of serum may be attributed to the variety of functions which the serum supplement may serve. It may provide both nutrient factors and regulatory factors such as hormones: These may be macromolecules or low-molecular-weight molecules, and the latter may be either free or bound to a macromolecular carrier. It is involved in the neutralization of trypsin used in subculturing and in the repair of cell damage caused by subculturing. It may sequester toxic substances such as heavy metals. It may bind, and release as required, nutrients which are insoluble, labile, or toxic at high concentrations, and may provide a carrier function for the utilization of others. It may also bind to the charged groups on the substratum and modify it to promote cell attachment. Different batches of serum can vary quite widely in composition and in growth-promoting ability. Fetal calf serum is widely used because of its high success rate but it is expensive, and provided one is prepared to screen batches for their suitability for the growth of specified cell types the cheaper newborn calf serum is often highly satisfactory.

The use of dialyzed serum provides a means of examining the requirement for low-molecular-weight growth factors. However, small molecules may remain bound to macromolecules or be generated in dialyzed serum by enzymic action, as in the case of amino acids produced by the action of endogenous proteases on serum protein. Lipids, which remain bound to apo-lipoproteins and serum albumin in dialyzed serum, can be removed by extraction with organic solvents (37,38).

Sera for cell culture are commonly heated to 56°C for 30 minutes prior to use to destroy complement which may lyse cells if the serum contains antibody to cell-surface antigens (Section 2.1.5). This treatment may also have the advantage of inactivating some serum-derived enzyme activities. However, it is not always essential and may be detrimental to cells requiring heat-labile growth factors.

1.1.9. Serum-Free Media

While chemically defined media have been available since the late 1950s which will support serum-free growth of lines, such as mouse L cells or HeLa cells, already subjected to strong selection for *in vitro* growth, all require supplementation with serum for growth of normal diploid cells (39,40). The strategy necessary for the development of media for completely serum-free growth of diploid cells is reviewed by Ham (40). Medium MDCB105 (Table 1.1) has thus been optimized for the growth and cloning of normal human fibroblasts with minimal serum supplementation, provided in the form of dialyzed fetal calf serum protein. In the absence of serum protein, if appropriate subculture techniques and substrata are used, single diploid fibroblasts will attach and spread in MDCB105 and will remain viable indefinitely, resuming replication when dialyzed serum protein is added, suggesting that the function of serum is limited here to the provision of stimuli for multiplication. The serum-derived activity responsible has been purified 15-fold by fractional precipitation, but the active preparation obtained is still very heterogeneous (41). Sato and collaborators (42,43) have succeeded in obtaining serum-free growth of a number of cell types, including primary cultures, in F12 or a mixture of F12 with Dulbecco's modification of Eagle's medium, supplemented with purified hormones and growth factors. Table 1.2 lists the supplements required for three cell lines. Each cell line has its unique requirements but in general they include insulin, transferrin, and a hormone localizing in the cell nucleus.

Iscove and Melchers (44) have achieved serum-free suspension growth of B lymphocytes in medium supplemented with serum albumin, transferrin, and soybean lipid. Modifications of this medium have been used to obtain serum-free growth of myeloid leukemia cells (45) and T and B lymphoid cell lines (46) in suspension culture.

Water purity (19) is even more crucial for serum-free culture than for culture in the presence of serum, because of the ability of serum to sequester and detoxify contaminant molecules. In addition subculturing techniques and substrata (Sections 1.2 and 1.4) have considerable influence on the success of serum-free culture of anchorage-dependent cells.

TABLE 1.2.   HUMORAL REQUIREMENTS OF CELL LINES IN SERUM-
FREE MEDIUM (42)

| | Cell Line | | |
Components	Melanoma M2R (g/L)	HeLa (g/L)	GH$_3$ Rat Pituitary Cell Line (g/L)
Insulin	5×10^{-3}	5×10^{-3}	5×10^{-3}
Transferrin	5×10^{-3}	5×10^{-3}	5×10^{-3}
Progesterone	6×10^{-6}		
Hydrocortisone		1×10^{-5}	
3,3′5′-Triiodothyronine			2×10^{-8}
Lutenizing hormone releasing hormone	1×10^{-5}		
Follicle stimulating hormone	4×10^{-4}		
Nerve growth factor	1×10^{-6}		
Epidermal growth factor		1×10^{-6}	
Fibroblast growth factor		1×10^{-4}	1×10^{-6}
Thyrotropin releasing hormone			1×10^{-6}
Parathyroid hormone			5×10^{-7}
Somatomedin C			1×10^{-6}

1.2.   THE SUBSTRATUM

The surfaces most commonly employed for growth of anchorage-dependent cells are glass and surface-treated polystyrene. Some formulations of both soda and borosilicate glass release toxic material into the medium and therefore the brand used must be empirically chosen for low toxicity (7). Critical parameters for optimal cell growth are the density of negative charges on the surface of the glass and its sodium content (47). Polystyrene surfaces can be given the required surface charge density by a number of techniques including treatment with concentrated sulfuric acid (7). Calcium ions play a role in attachment (47), presumably by bridging between the negative charges of the cell surface and the substratum. The surface of the substratum also becomes modified both by the cells, which produce a microexudate (48), and by the serum, from which fibronectin (cold-insoluble globulin,

LETS protein) is deposited on the surface (49). The importance of this modification is shown by the fact that many established cell lines require addition of fibronectin to serum-free media for attachment to occur: Normal fibroblasts, however, are capable of producing sufficient fibronectin for attachment (50). Coating of surfaces with a collagen gel (51) facilitates the attachment of many epithelial cell types: This appears to require fibronectin in some cases (52) but not in others (53). For some cell types exposure of the surface to a gelatin solution gives satisfactory results (54). For cloning of diploid fibroblasts at low serum concentrations, treatment of the culture surface with polylysine has been found beneficial (55,56). Van Wesel (57) was able to achieve a large surface area for cell growth in a manageable volume of medium by allowing cells to attach to beads of DEAE-Sephadex® A50 and maintaining them in suspension in spinner culture. Subsequent work has optimized the surface characteristics required for such microcarriers, and they are now commercially available in a variety of materials including surface-treated plastic (Lux Cytospheres^{TM}, Nunc Biosilon®), DEAE dextran (Flow Superbeads^{TM}, Pharmacia Cytodex®1), dextran derivatized with N,N,N, trimethyl 2-hydroxy aminopropyl groups on the bead surface only (Pharmacia Cytodex®2), dextran with a covalently bound collagen coating (Pharmacia Cytodex®3), and dimethylaminopropylacrylamide (BioRad Bio-Carriers^{TM}). Recent developments include medium perfusion to increase cell yield (58) and the use of fluorocarbon emulsions in which the interface with the growth medium is stabilized by adding polylysine to stabilize a bimolecular serum protein layer (59,60). In the latter system cells can be harvested without trypsinization by centrifuging to break the emulsion.

1.3. ESTABLISHMENT OF PRIMARY CULTURES

Techniques for the establishment of primary cultures are reviewed in references 7 and 61. Solid tissues may be either explanted as intact fragments or dissociated with a proteolytic enzyme. This is most commonly trypsin, but collagenase, papain, pronase, nagarse (7,61), and dispase (62) have all been used. A common problem in establishing primary cultures of epithelial cell types is overgrowth by connective tissue cells present in the starting tissue. They are often loosely described as "fibroblasts." However, such cells obtained from a variety of mouse organs have been found to resemble in their ultrastructure endothelial cells and pericytes from the capillary system and cells with morphologies intermediate between these two (63). This suggests that it is vasoformative mesenchymal cells rather than true fibroblasts which overgrow primary cultures, perhaps not a surprising result in view of their importance in wound healing. It seems best to reserve the term "fibroblast" for cells known to produce collagen and designate the above cells as "fibroblast-like" (64).

Likewise the term "epithelial" is used loosely in tissue-culture literature to describe any cell type growing in contiguous pavement-like sheets. Strictly it should be reserved for cells lining glands or surfaces other than those of blood and lymphatic vessels (endothelial) or coelomic cavities (mesothelial). Endothelial and mesothelial cells, despite their morphology, are of mesenchymal origin.

A number of techniques have been used to overcome the problem of overgrowth by fibroblast-like cells. Differential rates of attachment to the substratum can form the basis of an enrichment technique (65), as can differential rates of shedding from a solid explant cultured on a mesh grid (66). Some epithelial cell types can utilize D-valine as a source of L-valine by virtue of the enzyme D amino-acid oxidase which is absent from fibroblast-like cells (67). Polyamines are selectively toxic to fibroblast-like cells from a number of sources (68–70). An approach that has met with considerable success is to use serum-free hormone-supplemented media specifically tailored to the requirements of the epithelial cell type (71–74).

1.4. SUBCULTURING OF ANCHORAGE-DEPENDENT CELLS

Some epithelial cell types can be detached from the substratum mechanically, simply by pipetting, but in general detachment requires the use of trypsin, or one of the other proteolytic enzymes listed in Section 1.3 together with a chelating agent for divalent cations. Citrate was used as a chelating agent in early work but has been superseded by EDTA. Efficient dissociation with less cellular trauma than that caused by EDTA can be achieved with EGTA (ethyleneglycol-bis-(β-aminoethyl ether) N,N' tetraacetic acid) which specifically chelates calcium ions (75). The action of trypsin is stopped by the addition of serum: Most sera contain trypsin inhibitors, notably α-1-antitrypsin, but chicken serum is an exception and can be included in the dispersing solution to increase cell viability (76). When trypsin-dispersed cells are to be seeded into serum-free media, soybean trypsin inhibitor is commonly added to prevent further action of the trypsin (42). Hodges et al. (77) found that at room temperature trypsin was internalized by cells even in the presence of serum and remained detectable for 48 hr inside the cell. This effect did not occur at 4°C and McKeehan et al. (55) report increased viability of diploid human fibroblasts in media containing low concentrations of serum protein when the temperature of trypsinization is reduced to 4°C.

1.5. LIFESPAN OF DIPLOID CELLS IN CULTURE AND THE ESTABLISHMENT OF PERMANENT CELL LINES

Diploid human fibroblasts can be serially propagated in culture for only a limited number of passages (78). A similar phenomenon is observed with fibroblasts of other species although the situation is often complicated by overgrowth of diploid cells by aneuploid cells with unlimited growth potential (reviewed in ref. 79). The phenomenon responsible for the limited lifespan is termed senescence: Its exact relevance to aging of the organism has been the subject of considerable debate, although some relevance is indicated by the observations that culture lifespan correlates inversely with donor age within a species (79), and, in a study of eight mammalian species, was found to correlate with the square root of the maximum lifespan of the species (80). The lifespan *in vitro* of diploid cells other than fibroblasts has been little studied, although human diploid keratinocytes show a lifespan similar to that of fibroblasts, and possibly inversely correlated with donor age (81). The phenomenon by which cells with unlimited growth potential arise from diploid cells is properly termed "culture alteration" (64). For reasons that are not understood it is frequent in mouse cells, but relatively rare in human cells.

Holliday and collaborators have proposed a commitment theory to account for the finite lifespan of diploid cells (82). This theory postulates that, beginning with a population of potentially immortal cells, there is a given probability P that cell division will give rise to cells irreversibly committed to senescence and death. These cells initially multiply normally, but after a fixed number of cell divisions M all the descendants of each original committed cell die out. If P and M are sufficiently high, the number of uncommitted cells in the population will progressively decline to a point where, with normal subculturing regimes, they will inevitably be lost, leaving a mortal population. The finite lifespan, therefore, is a property not of individual cells, but of the progressively sampled population. It should be emphasized that this theory makes no assumptions about the mechanism of senescence and is compatible with a number of mechanisms (79). It is not my purpose here to describe these but rather to discuss the relationship between diploid, mortal cells and cells of established lines, and the importance of the theory for our purposes lies in its provision of a framework for doing so: By a reduction in P or in M a dynamic steady state can be produced in which sufficient uncommitted cells are produced by mitosis to make the probability of their loss by sampling remote, and the population is then immortal. The difference between mortal and immortal cell populations is thus a quantitative rather than a qualitative one. The theory in its simplest form fits well with experiment in a general way (82) although modifications have been proposed to allow for discrepancies in detail (83).

It is also possible to produce immortal cell lines from diploid cells by transformation with a tumor virus. However, it is important to realize that transformation and immortalization are distinct events. This has been clearly demonstrated with SV40-transformed human fibroblasts in which the "crisis" of senescence is merely delayed rather than abolished, and only as a rare event is it possible to derive cell lines of unlimited lifespan from senescent cultures (84,85). In general, isolation of stable SV40 transformants is a straightforward process only in species whose cells can readily be established as immortal lines even in the absence of virus. Cells with properties similar to virally transformed cells can also be produced by the introduction of exogenous DNA by the techniques described in Chapter 5.

An alternative approach is to immortalize the genome of a cell line with limited lifespan by constructing a fusion hybrid with an immortal cell line. This will be discussed in Chapter 4.

Table 1.3 lists some of the established cell lines most commonly used in somatic cell genetics. All are fibroblast-like lines as defined in Section 1.3. The L cell was one of the first cell lines to be established and its derivative L-929 was the first clonal cell line to be isolated. Unfortunately numerous sublines of the L cell have arisen as a result of independent multiplication by different laboratories during the 40 years since its isolation and the precise relationships between them are not in all cases clear, so that the possibility of variability between the stocks carried by different investigators must be particularly borne in mind for L cell derivatives. Confusion may also arise from the use of the designation "3T3:" This was

TABLE 1.3. SOME FIBROBLAST-LIKE ESTABLISHED CELL LINES COMMONLY USED IN SOMATIC CELL GENETICS

Cell Line	Origin	Modal Chromosome Number	Species Diploid Chromosome Number	Reference
L	Normal subcutaneous areolar and adipose tissue of adult male C3H mouse explanted into medium containing 20-methylcholanthrene	ND[a]	40	87
L929 (or NCTC-929)	Single-cell clone of L cells at passage 95	66[b]	40	88
LM	Derivative of L929 adpated for serum-free growth	62[b]	40	89
Balb/3T3	14–17-day mouse embryos	68[c]	40	90
CHO	Ovary biopsy from adult Chinese hamster	22[d]	22	91
CHO-K1	Proline-requiring subclone of CHO	20[b,d]	22	92
BHK-21	Kidneys of five unsexed 1-day Syrian hamsters	44[d]	44	93
BHK-21 (C13)	Single-cell clone of BHK-21	44[b]	44	93

[a] Not determined.
[b] Refers to seed stock held by American Type Culture Collection (ATCC), Rockville, Maryland.
[c] Refers to seed stock of subclone A31 held by ATCC.
[d] Taken from original publication.

used by Todaro and Green (86) to refer both to a subculturing regime used to establish cell lines (subculture every 3 days at 3×10^5 cells per 50-mm diameter dish) and to a cell line thus established from random-bred Swiss albino mice. The same regime has been used to establish cell lines from other mouse strains, designated, for example, Balb/3T3. To avoid ambiguity the names of these cell lines should be given in full and the original line designated 3T3-Swiss albino. Similar considerations apply to 3T6 and 3T12.

1.6. CHROMOSOME CONSTITUTION OF CULTURED CELLS

Before the introduction of chromosome banding techniques in the early 1970s, karyotype analysis was based predominantly on chromosome size and centromere position. However, a number of banding techniques are now available (94) which make it possible to distinguish all the chromosomes of a variety of mammalian species. Q banding is based on differential staining with quinacrine mustard or quinacrine dihydrochloride and examination by fluorescence microscopy. G banding, produced by Giemsa staining after (a) treatment with a proteolytic enzyme such as trypsin, (b) exposure to mild alkaline conditions, or (c) heating to 65°C in saline citrate, gives banding which corresponds exactly to Q banding with only a few exceptions. The relative usefulness of G and Q banding in a particular study is mainly determined by the fact that G banding is permanent, whereas Q banding is not and allows reanalysis of the same spreads by a second technique if desired. Banding patterns which are the inverse of Q and G bands (R bands) can be obtained by staining with either acridine orange or Giemsa after partial heat denaturation. The mechanistic basis of G, Q, and R banding was elucidated by the work of Dutrillaux (95) who found that, by allowing cells to incorporate BUdR at different phases of the growth cycle and then staining metaphase spreads with acridine orange, he could produce R, Q, or intermediate banding by varying the timing of exposure to BUdR. The patterns obtained indicate that R bands correspond to chromosome segments that replicate early in S phase and that Q bands correspond to those that replicate late. This technique has since been refined to produce permanent preparations by use of the photosensitive dye Hoechst 33258 to enhance photolysis of BUdR-substituted chromatin resulting in reduced affinity for Giemsa (96) and is compatible with *in situ* nucleic acid hybridization (Section

4.4). With all these techniques the resolution of the banding can be increased by preparation of metaphase spreads with chromosomes in a more extended configuration, either by the use of methotrexate synchronization followed by brief colcemid treatment (97) or by inducing premature chromosome condensation (Section 4.2). In the case of the BUdR/Hoescht 33258/Giemsa technique the BUdR itself can be made to perform the function of synchronization (98).

Different information is obtained by so-called C banding, which stains constitutive heterochromatin. Alkali- or formamide-induced denaturation is followed by incubation in saline citrate at 60°C to allow reassociation of repetitive sequences. The preparation is then stained with Giemsa. Dark staining is restricted to centromeric regions and secondary constrictions. The staining is particularly strong at the centromeres of mouse chromosomes, and is useful for their identification in interspecific hybrids. C-band size polymorphisms between different inbred mouse strains can also provide markers in intraspecific crosses (94).

A technique that distinguishes mouse from human chromosomes along the whole of their length is the use of Giemsa at pH 11. With this stain mouse chromosomes appear magenta and human chromosomes blue: the mechanistic basis is unknown. The technique is useful for detecting translocation between human and mouse chromosomes in fusion hybrids (94).

Active nucleolar organizer regions (NORs), the sites of multiple copies of ribosomal RNA genes which are visible as secondary constrictions, can be selectively stained using a silver nitrate-based technique. In fusion hybrids where the rRNA of one parent only is synthesized the NORs of the other parent remain unstained. It appears to be protein rather than DNA which is the basis of the staining (94). Other genes, including single-copy genes, can be localized by *in situ* hybridization (Section 4.4).

Karyotypic analysis is of great importance in monitoring the changes in chromosome constitution which occur during culture alteration and the selection of variants (e.g., Section 3.5.4), in defining the constitution of fusion hybrids (Sections 4.3 and 4.4), and in monitoring cross-contamination of cell lines. Culture alteration in mouse cells tends to give rise to a subtetraploid karyotype (90) whereas in other species, such as rat and rabbit (79), the change in karyotype is often more subtle and detectable only by banding techniques. The selective advantage enjoyed by aneuploid cells in culture is readily rationalized: The diploid chromosome constitution is the result of the action of natural selection on the organism as a whole where highly complex demands must be met. In contrast, the selective pressure in tissue culture simply operates in favor of the fastest-growing cell: It would indeed be surprising if these very different selective pressures favored cells of the same chromosome constitution. It should be emphasized that in aneuploid lines there is not only deviation from the diploid karyotype, but also heterogeneity of karyotype within one population and karyotypic evolution as selection proceeds. Therefore cell lines are characterized by a modal chromosome number, also known as the stem line number and designated $1s$, which is usually hyperdiploid. Many populations contain some cells with twice this basic number, designated $2s$.

1.7. DEGREE OF DIFFERENTIATION IN ESTABLISHED CELL LINES

Following Ephrussi (99) it is convenient to distinguish between ubiquitous or "household" metabolic functions essential for the maintenance and growth of any cell, and differentiated or "luxury" functions expressed only by certain cell types which, while necessary for the survival of the multicellular organism, are not essential for the survival of the cell. These categories have been alternatively designated "constitutive" and "facultative" markers, respectively (100). Most established cell lines show similar levels of expression of household functions to their tissue of origin but markedly reduced levels of expression of luxury functions. In most cases cultures are initiated from a heterogeneous mixture of cell types and selective overgrowth by connective tissue cells (Section 1.3) can readily account for the absence of expression of luxury functions characteristic of other cell types. However, this is not the whole story, since, for example, loss of expression of differentiated functions may also occur with passage in clonally derived cultures of liver Kupffer cells (101). Despite considerable success in maintaining the expression of differentiated functions in primary cultures and diploid cell lines of limited lifespan derived from normal tissues, in only a limited number of cases has it been possible to establish from them differentiated immortal cell lines either spontaneously (Table 1.4) or by chemical or viral transformation (Table 1.5). Although in some instances expression of differentiated functions has been shown to persist after SV40 transformation (84) the expression of such functions is often reduced or abolished on transformation (102,103), in some instances being lost before any karyotypic

TABLE 1.4. ESTABLISHED CELL LINES WHICH WERE DERIVED FROM NORMAL TISSUE WITHOUT KNOWN EXPOSURE TO AN ONCOGENIC AGENT AND WHICH EXPRESS DIFFERENTIATED FUNCTIONS

Cell Line	Source	Cell Type	Differentiated Function	Reference
3T3-L1	Mouse embryo	Preadipocyte	Lipid accumulation	112
G8	Fetal mouse limb muscle	Skeletal muscle myoblast	Myotube formation Acetylcholine sensitivity	113
H9c2(2-1)	Rat embryonic heart	Myoblast (skeletal-like)	Myotube formation Acetylcholine sensitivity	114
A7r5, A10	Rat embryonic thoracic aorta	Smooth muscle	Muscle-type CPK[a] Excitable membrane	115
BEcl-11, BEcl-17	Fetal rat brain	Glial cell	Surface antigens	116
FLC	Fetal mouse liver	Hepatocyte	Several including OTC[b]	117,118
MDCK	Canine kidney	Distal tubule epithelium	Structural and functional polarity	119
TM3	Fetal mouse testis	Leydig cell	Several including steroidogenesis	120

TM4	Fetal mouse testis	Sertoli cell	Several including secretion of H-Y	120
Several	Mouse	T lymphocyte	T-killer or T-helper function	14
BM1, BM2	Mouse bone marrow	Basophil/mast cell	Chloroacetate esterase Histamine production IgE receptors	121
Several	Human	T lymphocyte	T-killer function	14
RPM	Rat bone marrow	Promegakaryocyte	Several including Factor VIII antigen synthesis	122
C3	Adult rat calvaria	Osteogenic mesenchyme	Calcification	123
MC3T3-E1	Newborn mouse calvaria	Osteogenic mesenchyme	Calcification	124
Not named	Mouse placenta	Trophoblast	Secretion of gonadotropin, estradiol, progesterone	125
Several	Mouse embryo inner cell mass	Embryonal stem cell ("EK" cell)	Developmental pluripotency	See Chapter 6

[a] Creatine phosphokinase.
[b] Ornithine transcarbamylase.

TABLE 1.5. ESTABLISHED CELL LINES WHICH WERE DERIVED FROM NORMAL TISSUE TREATED *IN VITRO* WITH AN ONCOGENIC AGENT AND WHICH EXPRESS DIFFERENTIATED FUNCTIONS

Cell Line	Origin	Oncogenic Agent	Cell Type	Differentiated Function	Reference
L6	Newborn rat thigh muscle	Methylcholanthrene	Skeletal myoblast	Myotube formation	65
TPA30-1 TPA30-6 TPA209-9	Human placenta	SV40 tsA mutants	Placental	Alkaline phosphatase[a] hCG α-subunit[a,b]	126
RLA209-15	Rat fetal liver	SV40 tsA mutant	Hepatocyte	Albumin and transferrin synthesis[a] AFP synthesis[a,c] glucagon receptor	127
HE209	Human foreskin epidermis	SV40 tsA mutant early region DNA fragment	Keratinocyte	Keratins[a]	128
WC-5	Rat cerebellum	RSV[d] mutant ts for transformation	Astrocyte	GFAP[a,e]	129
HIT	Syrian hamster pancreatic islet	EMS[f] + SV40	β-cell	Insulin synthesis	130
Not named	Mouse fetal liver	A-MuLV[g]	B-lymphoid	Surface immunoglobulin	131
Many	Human	EBV[h]	B-lymphoid	Surface immunoglobulin	14
G1	Mouse fetal liver	FLV-A[i]	Erythroid precursor	Inducible hemoglobin synthesis	132

3BM-77, 3BM-78	Mouse bone marrow	FLV-P[j]	Erythroid precursor	Inducible hemoglobin synthesis	132
FLVA 16-IV	Mouse bone marrow	FLV-A[i]	Promyelocyte	Lysozyme; morphology promyelocyte + mature granulocyte	133
A(M) III	Mouse bone marrow	A-MuLV(M-MuLV)[k]	Promyelocyte	Lysozyme; morphology promyelocyte + mature granulocyte	133
427E	Mouse bone marrow	FLV-A[i]	Myelomonocytic	Promyelocyte morphology maturing in tumours	134
416B	Mouse bone marrow	FLV-P[j]	Bipotential hematopoietic cell	Morphological maturation to granulocyte + megakaryocyte See also footnote l	135

[a] Expression at the nonpermissive temperature only.
[b] Human chorionic gonadotropin.
[c] α-Fetoprotein.
[d] Rous sarcoma virus.
[e] Glial fibrillary acidic protein.
[f] Ethyl methanesulfonate.
[g] Abelson murine leukemia virus.
[h] Epstein–Barr virus (endogenous or exogenous).
[i] Friend leukemia virus, anemia strain.
[j] Friend leukemia virus, polycythemia strain.
[k] Abelson murine leukemia virus with helper virus coat.
[l] When cells are injected together with red blood cells into irradiated mice, survival is prolonged.

abnormality is detectable (102). The problem can be overcome by the use of viral mutants temperature-sensitive for the maintenance of transformation to obtain cell lines that grow progressively at the permissive temperature without expression of differentiated function, but that express differentiated function when shifted up to the nonpermissive temperature. Such lines were first obtained by transforming avian cells with temperature-sensitive avian sarcoma virus mutants (104), and both the avian virus mutants and tsA mutants of SV40 have now been used successfully with mammalian cells (Table 1.5). A promising strategy to facilitate the isolation of cell lines from normal tissue without the use of transforming viruses is to induce proliferation of the tissue before explanting. For instance, intracerebral injection of kainic acid produces a focal killing of neurons and a consequent proliferation of glial cells. If the gliotic areas are introduced into culture, cell lines are readily established (105).

Nevertheless, of the established cell lines currently available which express differentiated functions, the majority are of tumor origin (Tables 1.6–1.9). This is a consequence of the relative ease with which tumor cells can be established directly in culture as permanent lines without undergoing a crisis of senescence. Tumors very widely in their degree of differentiation, which in general shows a roughly inverse correlation with growth rate as illustrated by the Morris hepatomas (106). Nevertheless, by an appropriate choice of tumor it is usually possible to obtain immortal cell lines that express a chosen differentiated function. In some tissues tumors can be specifically induced using organotropic chemical carcinogens, such as the alkylnitrosoureas (107), or organotropic viral carcinogens, such as Abelson leukemia virus (14). When tumors are explanted into culture, contamination with fibroblast-like cells again presents a problem, but Buonassisi et al. (108) found that this could be overcome by alternately passaging the cells *in vitro* and *in vivo*. Passaging the cells as tumors appeared to enhance expression of differentiated function and even to reverse a loss experienced *in vitro*. The technique probably works by selecting *in vitro* in favor of culture-adapted variants of the tumor cells at the expense of the bulk of the tumor cells, and *in vivo* in favor of tumor cells at the expense of fibroblast-like cells, so that the cells which populate the secondary and succeeding tumors become progressively more capable of outgrowing the fibroblast-like contamination. It has now been used successfully for a number of tumors (109). A further source of differentiated cell lines is from differentiating cultures of teratocarcinoma cells. This will be discussed in Section 6.1.2.

TABLE 1.6.   ESTABLISHED HORMONE-PRODUCING CELL LINES
DERIVED FROM ENDOCRINE TUMORS

Cell Line	Tumor of Origin	Differentiated Function	Reference
GH1, GH3	Rat pituitary tumor	Growth hormone synthesis Prolactin synthesis (GH3 only)	136, 137
AtT20	Mouse pituitary tumor	ACTH[a] synthesis	138
rMTC 6-23	Rat medullary thyroid carcinoma	Neurotensin, calcitonin synthesis	139
Y1	Mouse adrenocortical tumor	Steroidogenesis	109
31A	Rat ovarian tumor (ovary transplant to spleen)	Steroidogenesis	140
BeWo	Human gestational choriocarcinoma	hCG[b] synthesis Conversion of pregnenolone to progesterone, estrone, and estradiol	63, 141
I-10	Mouse testicular Leydig cell tumor	Steroidogenesis	142
R2c	Rat testicular Leydig cell tumor	Steroidogenesis	143

[a] Adrenocorticotropic hormone.

[b] Human chorionic gonadotropin.

   The expression of differentiated functions may be modulated by addition
to the culture medium of diverse compounds. Some, such as hormones and
cyclic AMP, are physiologically relevant; others, such as bromodeoxyuridine,
are not but may provide important tools for the analysis of the mechanism
of differentiation (110,111). Reversible inhibition of expression of differentiated function is observed in a variety of cell types at concentrations of
bromodeoxyuridine which do not affect cell viability, but remarkably little
is known about its mechanism of action. While this is assumed to be mediated
by incorporation into DNA, the only established requirement is for conversion
to the monophosphate (111).

TABLE 1.7.   NEURAL TUMOR-DERIVED ESTABLISHED CELL LINES
WHICH EXPRESS DIFFERENTIATED FUNCTIONS

Cell Line	Tumor of Origin	Differentiated Function	Reference
Many	Mouse neuroblastoma c1300	Generation of action potentials Neurotransmitter synthesis	144,145
Several, e.g., IMR-32	Human neuroblastoma	Neurotransmitter synthesis	145,146
Several, e.g., B-104	Rat neuroblastoma	Generation of action potentials Neurotransmitter synthesis	145
PC12	Rat pheochromocytoma	Synthesis of noradrenaline, dopamine	146
C6	Rat glioma	S-100[a] synthesis	147
G26-20 and sibs	Mouse glioma	Several including S-100[a] and NS-1[b] synthesis	107,148
CHB	Human glioma	S-100[a], ganglioside synthesis	63,149
RN-2	Rat peripheral neurinoma (Schwann cell tumor)	S-100[a] synthesis CNP[c]	150

[a] Nervous system-specific acidic protein.
[b] Glial cell-specific surface antigen.
[c] 2',3' cyclic nucleotide 3' phosphohydrolase.

1.8.   CELL INTERACTIONS IN CULTURE

Cells in culture are subject to two distinct influences of cell density on growth rate. At low cell densities, growth rate commonly increases with cell density. A number of different mechanisms contribute to this "feeder effect" (see 168 and 169 for references), including both an effect due to contact with other cells or extracellular matrix secreted by them and a

TABLE 1.8.   HEMATOPOIETIC TUMOR-DERIVED ESTABLISHED CELL LINES WHICH EXPRESS DIFFERENTIATED FUNCTIONS

Tumor of Origin	Cell Line	Differentiated Function	Reference
Mouse B cell tumors including myelomas	Many at different stages of differentiation	Cytoplasmic, surface or secreted immunoglobulin	14
Human B cell tumors	Many at different stages of differentiation	Cytoplasmic, surface or secreted immunoglobulin	14
Mouse T lymphomas	Several	Several including expression of T-lymphocyte surface antigens	14
Human T-type acute lymphoblastic leukemia	Several	Several including expression of T-lymphocyte surface antigens	14
Mouse erythroleukemia (FLV-A[a] or FLV-P[b] induced)	Several, e.g., 745	Inducible synthesis of hemoglobin and other red cell products	132
Human chronic myeloid leukemia	K-562	Inducible synthesis of hemoglobin (note erythroid rather than myeloid properties)	132
Human promyelocytic leukemia	HL-60	Chloroacetate esterase Myeloperoxidase Inducible morphological maturation and phagocytic capacity	151
Mouse myeloid leukemia	M-1	Myeloblast morphology Inducible morphological maturation	152

[a] Friend leukemia virus, anemia strain.

[b] Friend leukemia virus, polycythemia strain.

TABLE 1.9. MISCELLANEOUS TUMOR-DERIVED ESTABLISHED CELL LINES WHICH EXPRESS DIFFERENTIATED FUNCTIONS

Cell Line	Tumor of Origin	Differentiated Function	Reference
HTC	Rat hepatoma (Morris 7288c)	Glucocorticoid-inducible TAT[a]	63,153
MH_1C_1	Rat hepatoma (Morris 7795)	Synthesis of albumin, serum proteins, and 9th component of complement	63,154
$7800C_1$	Rat hepatoma (Morris 7800)	Several including urea cycle enzymes	155
H4IIEC3,Fu5	Rat hepatoma (Reuber H-35)	Many including glucocorticoid-induced TAT[a]	156,157
BW-1	Mouse hepatoma	Synthesis of albumin, α-fetoprotein, aldolase B	158,159
M-3	Mouse melanoma	Melanogenesis	142
RPMI3460	Hamster melanotic melanoma	Melanogenesis	160
HFH14,HFH18	Mouse melanoma B16	Melanogenesis	63,161
Not named	Human tricholemmoma (benign hair follicle tumor)	3D hair follicular differentiation	162
Rama 25	Rat mammary tumor	Inducible casein synthesis	163
229E	Mouse intramuscular sarcoma (MSV[b]-induced)	Myotube formation, CPK,[c] myokinase	164
P815	Mouse mastocytoma	5-hydroxytryptamine, histamine synthesis	165
AR4-2J	Rat pancreatic exocrine tumor	Synthesis of chymotrypsin mRNA	166,167
Several	Mouse embryonal carcinoma	Developmental pluripotency	See Chapter 6
Several	Human embryonal carcinoma	Developmental pluripotency	See Chapter 6

[a] Tyrosine aminotransferase.

[b] Murine sarcoma virus.

[c] Creatine phosphokinase.

separate effect of diffusible molecules. The latter may be specific growth factors or simple molecules such as nonessential amino acids which are readily lost from cells if there is an appreciable concentration gradient across the cell membrane. The feeder effect must be considered in optimizing conditions for clonal growth of cells (Section 1.9).

At high density, there is a density-dependent growth inhibition (170). Loss of density-dependent growth inhibition is frequently associated with the acquisition of tumorigenicity, although the correlation is not perfect (171). Stoker (172) found that a polyoma-transformed Syrian hamster cell line which failed to show density-dependent growth inhibition when cultured alone, was subject to inhibition by normal cells. However, this is the exception rather than the rule. Although all normal cells, cultured alone, exhibit density-dependent growth inhibition, in mixed culture even two normal cell lines may fail to show reciprocal inhibition (168). The use of the term "contact inhibition" for density-dependent growth inhibition is to be avoided as it leads to confusion with the phenomenon of contact inhibition of cell locomotion (173) which probably occurs by a totally different mechanism.

A further cell interaction phenomenon which must be taken into account by the somatic cell geneticist is metabolic cooperation (174). This is the exchange of small molecules (with molecular weights below a cut-off value in the region of 1000 for mammalian cells) between cells in contact via gap junctions. As a consequence, at cell densities high enough for contact to be frequent, cells pool their intermediary metabolites whereas macromolecules remain in the cell in which they are synthesized. This can modify the phenotype of cells in contact with those of a different genotype and must be taken into account in the design of selective procedures (Section 2.1.7).

1.9. CLONING

Genetic analysis of cells necessitates isolating the progeny of single cells and hence cloning techniques (7,175) are of paramount importance. The principal technical problem to be overcome in order to achieve high cloning efficiency is to compensate for the lack of feeder effect (Section 1.8) at low cell density. For some cell types acceptable cloning efficiencies can be achieved by supplementing the normal growth medium with nonessential amino acids and sodium pyruvate to overcome the loss of these metabolites from the cell. Where specific growth factors are important cloning may be carried out in medium in which cells have previously been allowed to grow at higher density and into which such factors have thereby been released ("conditioned" medium). In other cases where cell contact is important the cells to be cloned may be seeded onto a feeder layer of cells rendered incapable of cell division by X-irradiation or treatment with mitomycin C. With X-irradiated feeder layers the radiation dose must be precisely controlled to avoid contamination of the cloned cells by feeder cells which have escaped its effects. With mitomycin C the treatment is more conveniently standardized, but adverse effects on the cloned cells have been claimed (175). Where more than one colony arises within a culture vessel, single colonies can be isolated either by the use of coverslip fragments as a substratum (7) or with the aid of cloning rings (175). However, because cells can migrate between colonies it is preferable to clone by dilution, ensuring by microscopic observation that each culture contains only one cell. For dilution cloning it is advantageous to keep the volume of medium as small as possible to allow conditioning by the cell being cloned. This was originally achieved using microdrops of growth medium under liquid paraffin (176). However, some batches of liquid paraffin are toxic and it is now preferable to use multiwell culture trays specifically made for cloning in which each well accommodates a medium volume of 0.1–0.2 mL. A recent modification of this technique employs wells with a geometry designed to allow the use of phase-contrast microscopy (Optikon multiwell dishes, Northumbria Biologicals Ltd.) which facilitate verification that only a single cell is present.

Cloning of suspension cells may be carried out in media solidified by the addition of agar or agarose. This may also be used as a selective system for the isolation of virus-transformed derivatives of anchorage-dependent cells. Feeder cells can be provided if required by the use of a cell monolayer beneath the agarose layer. An alternative approach is to use nondividing suspension cells such as thymocytes ("filler" cells) to allow dilution cloning (44).

1.10. STORAGE OF CELLS BY FREEZING

Preservation of cell stocks by freezing is indispensable to the somatic cell geneticist for a number of reasons. Because the properties of cell lines undergo evolution with time it is valuable to be able to return to early-passage cells which have been previously characterized; the number of potentially variant cell lines generated in a single experiment may be too large to allow them all to be screened simultaneously, and freezing allows more convient scheduling; also, of course, freezing provides back-up in the event of loss or contamination of the growing stock. Cells are normally frozen in suspension in the normal growth medium supplemented with glycerol or DMSO as a cryoprotective agent (7,177). Other cryoprotective agents have recently been identified (178): Many are also effective inducers of differentiation in erythroleukemia cells and promotors of polyethylene glycol-mediated cell fusion (Chapter 4), suggesting that they exert their action by effects on the cell membrane. A minimum cell density is required for good survival of the suspension and, if small numbers of cells are to be frozen, filler cells can be added as for cloning of suspension cells (Section 1.9). Fragile cells may survive better in undiluted serum plus the cryoprotective agent. Storage at below $-130°C$ gives best results because the growth of ice crystals is retarded at these temperatures, and it is routine to maintain stocks in the vapor above liquid nitrogen. Slow freezing (approx. 1°C/min) and fast thawing give best results (for details see ref. 177).

2

SELECTIVE TECHNIQUES

Selective techniques are of central importance in somatic cell genetics because of the variety of situations in which a cell with a phenotype of interest is generated as a rare event in a population. They allow the isolation of mutants or other classes of variant generated spontaneously or by the deliberate application of a mutagen (Chapter 3). They are necessary for the isolation of viable fusion hybrids, as the formation of the viable hybrid from the initial fusion product, the heterokaryon, is a rare event (Chapter 4). They are also necessary in gene transfer by techniques such as transfection, where stable incorporation of the introduced genetic material into the host chromosome is a rare event (Chapter 5). While the majority of these techniques are based on selective survival, physical separation techniques are becoming increasingly important and will be discussed here too. Also included under this heading are a number of isolation techniques which in the strict sense of the word are not selective at all but are rapid screening techniques, namely, sib selection, "toothpick," and replica plating methods.

2.1. TECHNIQUES BASED ON SELECTIVE SURVIVAL

2.1.1. Toxic Antimetabolites

Selection for resistance to toxic antimetabolites was one of the earliest methods successfully applied to the isolation of variant cells, and has since been widely used. The best antimetabolites for this purpose, which in many cases are analogues of normal metabolites, are those that either exert their toxicity by acting upon a unique target or require activation by a unique metabolic pathway. Resistance may arise either by an alteration in the target or by a change that restricts access of the toxic species to the target, such as in permeability, metabolic activation, or degradation. Before the advent of high-efficiency cloning techniques selection for resistance was carried out by "training" procedures in which a starting culture was split into a number of subcultures which were then exposed to different concentrations of the antimetabolite. Because of the lack of feeder effect at low cell density this tended to produce either high survival or none at all. The culture exposed to the highest concentration of the antimetabolite allowing survival was then split and the process repeated, and during the course of a protracted multistep procedure it was possible to "train" cells to grow in progressively higher concentrations of antimetabolite. While the method was designed to produce progressive enrichment for stable mutant phenotypes, frequently in practice cell lines were obtained which had acquired resistance in a series of increments, often as a result of gene amplification, and whose resistance was unstable in the absence of continued selective pressure. The best understood example is that of methotrexate resistance (Section 3.5.4). Selections are still sometimes performed by the "training" method but now that it is a straightforward matter for most cell lines to obtain high cloning efficiencies it is preferable to select for the desired degree of resistance in a single step, using a concentration of toxic antimetabolite which will eliminate all the sensitive cells and allow resistant cells to form colonies that can then be expanded directly into resistant lines.

2.1.2. Radioactively Labeled Compounds

The use of radioactive compounds as selective agents depends upon the cytotoxic effect of the emitted radiation. Clearly one wishes this toxicity to be restricted as far as possible to the cell that has taken up the isotope, which limits the useful isotopes to those whose radiation has a path length that is short compared with the diameter of a cell. In practice tritium is the only isotope to have been extensively used. The most efficient killing is obtained by incorporation of ^{3}H-thymidine into DNA; progressively less efficient killing is obtained with ^{3}H-uridine incorporated into nuclear RNA, ^{3}H-uridine incorporated into cytoplasmic RNA, and ^{3}H-labeled amino acid in the acid-soluble fraction of the cytoplasm (179). The efficiency of killing can be increased by freezing the cells that have incorporated isotope, thus preventing diffusion or metabolism of the labeled compound while allowing disintegrations to be accumulated over a period of days or weeks (179,180). Depending upon the labeling conditions, one can select for resistance on the basis of either reduced membrane transport or reduced incorporation into macromolecules.

2.1.3. Selection for Auxotrophy and Prototrophy

Techniques available for the isolation of auxotrophs are adaptations of the penicillin-enrichment technique used in bacteria where cells are starved of the metabolite of interest and then exposed to an agent that is toxic only to growing cells, after which the surviving, nongrowing cells are rescued into complete medium. Agents used to kill growing cells selectively are ^{3}H-thymidine (Section 2.1.2), 5-bromodeoxyuridine (BUdR, BrdU), whose incorporation into DNA renders the latter sensitive to visible light (181), and 5-fluorodeoxyuridine (FUdR, FdU; ref. 182). The mode of killing by FUdR is not completely understood but it is known to inhibit thymidylate synthetase (183) and it is thought that once cells reach the G_1/S boundary in the cell cycle they undergo loss of viability by a phenomenon similar to the "thymineless death" or "unbalanced growth" seen in *E. coli* (184), while cells arrested by starvation in other parts of the cell cycle are unaffected. Uridine is normally added along with FUdR to minimize incorporation of breakdown products into RNA and consequent spurious killing.

Selection for prototrophy is comparatively simple: Auxotrophic cells are simply exposed to medium depleted of the metabolite which they require. However, in selections of both kinds it may not be a straightforward matter to achieve stringent starvation, and, in particular, in work with amino acids it has sometimes been necessary to use dialyzed serum and/or degradative enzymes to remove amino acids generated by serum proteolysis (182,185).

2.1.4. Selection for Conditional Lethal Variants

The techniques described above for the isolation of auxotrophs can readily be applied to the isolation of temperature-sensitive cell variants (186). Mammalian cells grow very slowly below 33°C and lose viability at temperatures in excess of 39°C, which constrains the choice of permissive and nonpermissive temperatures. One cycle of selection consists of shifting cells to the nonpermissive temperature for sufficient time to allow temperature-sensitive cells to cease growth and then exposing them at the nonpermissive temperature to an agent selectively toxic to growing cells (Section 2.1.3). This agent is then removed and the cells incubated at the permissive temperature to allow growth of the survivors.

2.1.5. Selective Survival Dependent Upon the Binding of Antibodies or Other Ligands to the Cell Surface

Two methods are available for selectively killing cells that are capable of binding antibody to surface antigens. The first is complement-mediated cytotoxicity (187). Complement, usually added in the form of rabbit or guinea pig serum, is activated by interaction with the Fc region of antibody–antigen complexes at the cell surface, and this triggers a cascade of proteolytic zymogens which leads finally to the lysis of the cell (188). However, typically about 1% of the target cells survive killing so that multiple rounds of selection are required (187), although in the selection of membrane IgM⁻ variants from lymphoma cells the efficiency of the kill can be improved by selecting twice within a cell cycle (189), and although isolation of an HLA variant from a human lymphoid line has been achieved in a single step (190). Furthermore, the level of toxicity depends upon features of the target cell other than antigen concentration (191) and many cultured cell lines are not susceptible (192). Nevertheless, with certain antigen–cell combinations, complement-mediated cytotoxicity gives very satisfactory results as a selective system and it has been used for the isolation of segregants from human–rodent somatic cell hybrids (193–195), and of a number of variants from lymphoid cells (189 and references cited therein).

A more recent development with wider potential for application in somatic cell genetics is that of immunotoxins (196,197). Here the specificity of an antibody is used to deliver a toxic molecule to a target cell carrying the cognate antigen. Plant toxins such as ricin and abrin have properties particularly favorable for this application. They are heterodimers of a toxic A polypeptide chain, which in the case of ricin inhibits the 60s ribosomal subunit, and a B polypeptide chain, which binds to the surface of the cell and mediates entry of the A chain into the cytoplasm. Antibodies can be chemically coupled to the whole toxin, to chemically derived A chains (ricin-A, abrin-A), or to naturally occurring toxins such as gelonin which correspond to the free A chain. In the case of antibodies coupled to the whole toxin, lactose or galactose can be used to prevent binding to the cell through the B subunit leaving only the antibody-specific binding; antibodies coupled to free A chains bind only to their cognate antigen. The coupling method used is critically important to avoid homopolymerization and inactivation of toxin or binding site, and to produce a bond readily cleaved to allow delivery of the A chain to the cytoplasm: The method of choice appears to be use of the reagent N-succinimidyl 3 (2-pyridyldithio) propionate (Pharmacia SPDP reagent) which produces a readily cleaved disulfide bond between A chain and antibody (198). An alternative is to use hybrid antibodies with one combining site for a toxin and one for a cell surface determinant. Such hybrid antibodies can be produced either by mild reduction and reoxidation of a mixture of $F(ab')_2$ fragments of two antibodies (199) or by using "hybrid hybridomas" (200). In general, immunotoxins containing divalent antibody fragments are more toxic than those containing monovalent fragments, showing that cross-linking of antigens is important for toxicity, probably in mediating endocytosis (196). Some monoclonal antibodies give rise to immunotoxins of poor toxicity but a number of techniques are available to overcome this problem (196). Immunotoxins have not yet been widely used in somatic cell genetics but their success in isolating subpopulations of cells from normal tissues shows that they have the required selectivity. Lectins and other ligands can be successfully substituted for the antibody molecule to widen the range of applications.

Two techniques allowing selection in favor of cells that bind a chosen antibody molecule have recently been described (201). In the first, the antibody is coupled to ferritin and added to a culture in serum-free medium lacking transferrin: As most cells lack receptors for ferritin only those cells that bind antibody can take up the iron required for growth. The efficacy of this selective system has been demonstrated in reconstruction selections using mixtures of PC12 pheochromocytoma cells and a variant lacking the PUNC surface antigen. In the second, selenium-containing monoclonal antibody is produced by growing hybridoma cells in medium containing selenomethionine in place of methionine and used as the only source of selenium for cultures in serum-free medium. This is shown to be capable of supplying the selenium requirement of cells binding the antibody, but no data are presented for cells that do not (201).

Antibody-based selection techniques involving physical cell separations are discussed in Section 2.2 and a nonselective antibody-based isolation technique in Section 2.3.1.

2.1.6. Irreversible Biochemical Inhibitors

Wright (202,203) has described a technique in which two cell populations treated respectively with the inhibitors iodoacetamide and diethylpyrocarbonate undergo mutual rescue on heterokaryon formation. This allows enrichment of the proportion of heterokaryons in fusogen-treated cell populations to about 99%, but because of the low efficiency with which viable hybrids are formed from heterokaryons the method is less useful for the isolation of viable hybrids.

2.1.7. Effect of Cell Interactions on Selective Survival

In the design of selective techniques for analogue resistance due account must be taken of the effects of cell interactions, which may modify the phenotype of wild-type or mutant cells. Good examples are to be found in selections involving nucleic acid precursors and their analogues. The purine analogue 6-thioguanine, for instance, is toxic to wild-type cells by virtue of being metabolized to the nucleotide, which is then incorporated into nucleic acid. Variant cells lacking the enzyme HGPRT (hypoxanthine guanine phosphoribosyl transferase) are resistant to 6-thioguanine because they cannot convert it to the nucleotide, but become sensitive on forming gap junctions with wild-type cells due to intercellular transfer of toxic nucleotide formed in the wild-type cell ("kiss of death," ref. 204) Thus isolation of HGPRT⁻ variants from a wild-type population becomes progressively more inefficient as the cell density is increased above that at which the number of cell contacts becomes appreciable, due to loss of potentially resistant cells. Conversely, intercellular transfer of molecules can rescue a cell from the effects of an otherwise toxic environment. In HAT medium (Section 3.4) the products of both HGPRT and thymidine kinase (TK) are required for growth. However, HGPRT⁻ and TK⁻ cells can be rescued, either by each other or by wild-type cells, via the passage of nucleotides from cell to cell through gap junctions in high density cultures. Poor enrichment of wild-type cells from HGPRT⁻ or TK⁻ populations is therefore obtained at high cell density due to inefficient killing of parental cells.

2.2. TECHNIQUES BASED ON PHYSICAL SEPARATION

2.2.1. Cell Affinity Chromatography

Cells carrying a chosen surface antigen are selectively bound by adsorbents to which the cognate antibody has been covalently linked, so that the unbound fraction of cells is enriched for those that do not carry the antigen (205). A similar approach may be employed with other ligands such as lectins. Because of the strength of interaction between antibody and antigen, recovery of the bound fraction of the cells in a viable state is more difficult to achieve. Two approaches which have successfully been used are enzymic degradation of the adsorbent matrix (206) and elution with excess antigen, which for the case in point was homologous IgG (207). Alternatively, the adsorbent may be derivatized with *Staphylococcus aureus* Protein A and the cells pretreated with antibody in solution prior to applying them to the column. Antigen-carrying cells bind antibody which then interacts through its Fc portion with the immobilized Protein A, and excess soluble IgG can then be used to achieve desorption without the necessity of dissociating the antigen–antibody complex (208). A particularly promising technique is the use of antibody-coated plastic dishes (209). Antibody molecules (and other proteins) will adsorb to the surface of bacteriological grade petri dishes without the necessity for a covalent cross-linking reagent and the dishes will then specifically bind cells carrying the cognate antigen. Tissue-culture grade dishes give inferior results as the amount of nonspecific binding is increased. Elution of the bound cells can then be achieved by gentle pipetting in saline containing 1% serum. One drawback of the method is that it requires affinity-purified or high-titre antibody, since other proteins compete for binding to the dishes. However, crude antibodies can be applied to the cells in solution and fractionation then achieved using an affinity-purified second antibody directed against the first. In this way a single purified antibody suffices for a wide range of different antigens. Enrichment factors of up to 100-fold are readily achieved and there seems no reason why much higher ones should not be possible particularly as successive rounds of selection, in principle, can be applied directly one after the other.

2.2.2. Rosetting

Instead of being immobilized on an adsorbent, antibodies may be coupled to erythrocytes (usually sheep or bovine) by the use of reagents such as chromic chloride. The coated erythrocytes will then form rosettes with cells carrying the corresponding antigen on their surface (210). Rosettes may be separated from nonrosetting cells by centrifugation after applying the suspension on the surface of a denser layer of Ficoll/sodium metrizoate (211). Both populations can be recovered with high viability, but the rosetted population is usually contaminated with antigen-negative cells. Therefore the method is best-suited for the isolation of the antigen-negative subpopulation.

2.2.3. Fluorescence-Activated Cell Sorting

In the fluorescence-activated cell sorter (FACS) a suspension of cells is passed, in a stream of droplets of such a size that very few contain more than one cell, through a laser beam. Detectors quantify low-angle light scatter and fluorescence emission at either one or two chosen wavelengths. Cells producing signals within a desired range for any combination of these parameters can be deflected electrically and separated from the remainder of the population (212). Low-angle light scatter essentially measures cell size, while fluorescence emission can be made to parallel any of a wide range of cellular parameters depending upon the fluorescent tracer employed. These parameters include: (a) DNA content, using fluorochromes such as Hoechst 33258 (213); (b) levels of a particular enzyme, for example, folate reductase, using the fluorescein conjugate of the irreversible inhibitor methotrexate (214); (c) ability to form gap junctions, using the tracer carboxyfluorescein diacetate (215); and, by far the most exploited, (d) expression of surface antigens, using antibodies conjugated to fluorescent dyes (216,217). The disadvantage of the technique, apart from the high cost of the apparatus, is that only about 10^7 cells per hour can be sorted, although if the desired cells constitute less than 5% of the population it is advantageous to carry out a higher-speed presort which produces a population enriched for the cells of interest but still appreciably contaminated with cells from the bulk of the population: This may then be resorted with greater discrimination (212). Techniques for FACS sorting of cell populations stained with conjugated antibodies are already highly sophisticated, with recent developments including (a) the use of antibodies coupled to fluorescent microspheres, which increases sensitivity and signal-to-noise ratio; (b) removal of dying cells, which stain nonspecifically, by use of propidium iodide which causes them to fluoresce in the red part of the spectrum, allowing them to be rejected; and (c) the use of directly conjugated primary antibodies to reduce nonspecific staining (217). Enrichment factors of 250-fold in a single round of sorting can be readily achieved. The FACS has been used to isolate myeloma and hybridoma variants which have undergone heavy-chain class switching (216,217). These variants, which occurred at frequencies between 10^{-5} and 10^{-7}, required between three and seven rounds of sorting to raise their proportion in the population to a level where they could be isolated by cloning—a step that can be achieved directly on the sorter using a recently developed attachment (217). A particularly interesting example of the use of the FACS to isolate variant cells is the selection of a variant from the lymphoma cell line LDHB with a structurally altered H2-K^k antigen (218). This used two monoclonal antibodies directed against separate epitopes of the H2-K^k antigen molecule which competitively inhibited each other's binding, presumably because of the close proximity of the two epitopes. Antibody A was conjugated to fluorescein, while antibody B was unconjugated and present in 1000-fold excess. The wild-type cells therefore exhibited no

staining. Variant cells that failed to express H2-K^k at all would also show
no staining. Only those variants with a structurally altered antigen in which
the binding site of antibody B was destroyed while that of A was retained
would fluoresce. Such cells arose at a frequency of 10^{-5}–10^{-6} and could
be isolated after three rounds of sorting.

2.3. NONSELECTIVE ISOLATION TECHNIQUES

Techniques have been developed which allow the isolation of any variant,
hybrid, transfectant, and so on whose phenotype is scorable by an assay
which can be performed *in situ* on single colonies. These are essentially
techniques which allow rapid screening of a large number of colonies: No
selective system is necessary, the only requirement being that the desired
phenotype should occur at a sufficiently high frequency for it to be feasible
to score sufficient colonies—in practice, 1×10^{-4} and above. Where the
screening technique can be carried out without killing the cells, colonies
with the desired phenotype can be picked from the screening plate. Where
it cannot, use can be made of the technique of sib selection, "toothpick"
methods, or replica-plating (Sections 2.3.2, 2.3.3, and 2.3.4).

2.3.1. Screening Techniques That Do Not Destroy Viability

Morphological variation can be used to identify colonies of interest, particularly in the case of somatic cell hybrids where resemblance to one or
other parent often correlates with chromosome balance (219). The lipid-
accumulating preadipocyte cell line 3T3-L1 was isolated by picking foci
of lipid-accumulating cells seen in the parent 3T3-Swiss albino line (112).
Maio and de Carli (220) devised a screening technique for the isolation of
alkaline phosphatase-defective variants by overlaying colonies with agar
containing the chromogenic substrate *p*-nitrophenyl phosphate. Scoring of
colonies was possible within 5 minutes and viable cells could be recovered
up to half an hour later. Thompson et al. (221) isolated variants sensitive
to UV, mitomycin, and ethyl methane sulfonate by treating with low doses
of the appropriate agent and picking colonies that showed impaired growth.
Mouse myeloma cell lines defective in immunoglobulin secretion have been
isolated by plating in soft agar, overlaying with antiserum to mouse immunoglobulin and scoring for the presence of a precipitate in the region of

the colony (222,223). Hybridoma colonies secreting an antibody of desired specificity can be isolated by applying a nitrocellulose filter precoated with antigen to the surface of an agarose culture, removing, and challenging the adsorbed antibody with labeled antigen. By precoating with antiimmunoglobulin in place of antigen, colonies secreting any immunoglobulin, or immunoglobulin of a desired class or subclass, can be identified (224). Colonies secreting an antibody of desired specificity can also be isolated using the hemolytic plaque assay (225,226), in which the cells are plated in agarose with erythrocytes to which the desired antigen has been coupled, and the plates are subsequently developed with complement in the presence or absence of antiimmunoglobulin ("developing antibody"). Colonies secreting IgM of the desired specificity in general will produce a hemolytic plaque in the absence of developing antibody, but if the antibody is of any other class the developing antibody is necessary for complement fixation and hence for plaque formation (225). A reverse hemolytic plaque assay has been developed by Feldman and Chou (227) which enables colonies to be screened for the secretion of specific polypeptides. Antibody directed against the polypeptide of interest is coupled to erythrocytes which are then applied to an agar overlay above the colonies. On development with complement and soluble antibody of the same specificity (facilitating antibody) plaques appear above colonies secreting the chosen polypeptide. The technique has been successfully used with α-fetoprotein, transferrin, and human chorionic gonadotrophin.

2.3.2.   Sib Selection

In this technique, originally developed for bacteria (228), one scores the frequency of a desired phenotype in a population by a screening method, and then sets up a number of subcultures from the original population. The number of cells used to seed each subculture is chosen so that a small number of the desired variant cells is distributed among a relatively large number of subcultures. After a short period of growth each subculture is split into two replicates, one replicate being scored for variant frequency and the other retained for subsequent growth. Assuming that variant cells grow at the same rate as the rest of the population, some subcultures should show an increased frequency of variants compared with the starting population whereas other subcultures should show no variants. The subculture with the highest variant frequency is chosen and the process is repeated. In this way a gradual enrichment is achieved and eventually the desired cells can be isolated by cloning. This method has been applied to demonstrate that segregants resistant to 6-thioguanine or BUdR arise from hybrids between HGPRT⁻ and TK⁻ Syrian hamster cells in the absence of selective pressure (229), and to isolate variant CHO cells defective in glucose 6-phosphate dehydrogenase (230). The latter variants were present in mutagen-treated populations at frequencies of approximately 6×10^{-4} and could be isolated after five or six rounds of enrichment. Sib selection has also been used to isolate CHO variants heterozygous at the adenine phosphoribosyl-transferase (*aprt*) locus (231). The method is laborious and requires extensive cell multiplication but is of value when technical difficulties preclude the use of methods described in the following sections.

2.3.3. "Toothpick" Techniques

In these techniques, again borrowed from microbial genetics, colonies growing on a master plate are transferred with a sterile toothpick or similar implement to gridded replicate plates, one or more of which then serve for screening and one for recovery of the corresponding colony. The method is readily applied to anchorage-independent cells by growing them on an agar surface and making transfers with a sterile glass rod (232). Its application to anchorage-dependent cells is more problematical because of the difficulty of removing them from the substrate, but Jeggo et al. (233) have described a technique in which CHO cells growing on a conventional substrate are overlaid with agar. The cells round up into the agar and can then be transferred with a toothpick onto the surface of gridded agar plates.

2.3.4. Replica Plating

A further improvement in the speed of screening is made possible by techniques in which a pattern of colonies can be directly transferred in a single operation from a master plate to a replica. As originally developed for bacteria (234) the replica-plating technique involved the transfer of colonies from a master agar plate to a velveteen pad and thence to a second agar plate. Adaptation of the technique to tissue-culture cells has been a long and difficult process. Early replica-plating devices were simply a means of making serial transfers from a large number of microwells simultaneously: Since each well required trypsinization prior to making transfers the method was extremely laborious (235). Attempts were made to apply the bacterial technique directly to anchorage-independent cells growing on agar surfaces, but problems were experienced with smearing because of the softer agar which must be used with mammalian cells and with poor fidelity of replication (232). Replicas could be produced by overlaying colonies of anchorage-dependent cells with nylon cloth, but the efficiency of transfer was only 80–90% and the formation of satellite colonies made interpretation difficult (236,237). The first routinely usable technique was developed by Esko and Raetz (238) and involved colony formation under a layer of filter paper weighed down with glass beads. After a period of growth the filter paper was removed and carried a replica of the pattern of colonies on the surface of the tissue-culture dish. Faithful replicas could be achieved with CHO cells by this technique and it has been used to isolate a number of variants (Section 3.4). A modification of the technique has recently been published (239) in which sheets of polyester mesh are substituted for the filter paper, and the plastic surface of the dish is coated with polylysine. This modification widens the range of cell lines from which faithful replicas can be achieved and it has been successfully used with hybridomas (239). It also enables one to produce multiple replicas by stacking several sheets of mesh.

2.4. OVERALL STRATEGY IN THE USE OF ISOLATION TECHNIQUES

With most selective systems it is impossible to choose conditions where one can achieve elimination of all wild-type cells and recovery of all variant cells in a single round of selection. (For simplicity I shall refer in this section to "wild-type" and "variant" cells, although the same considerations apply to any bulk population and the cells to be isolated from it.) It is therefore necessary to compromise so that one achieves enrichment for variant cells together with acceptable recovery. In order to isolate the cells of interest it may be necessary either to perform multiple rounds of selection using a single technique or to apply different techniques in succession. Where multiple rounds of selection using a single technique are to be used to isolate variants which occur at very low frequency in a population, it may well be advantageous to use different conditions in different rounds of selection. In the first round of selection the priority is to achieve high recovery of the variant cells even if the enrichment factor is relatively modest, otherwise one may eliminate all of them from the population. Once some enrichment has been achieved, one can then tolerate a lower recovery and may thereby achieve a higher enrichment factor (see, e.g., ref. 240). In selecting for autosomal variants it may be advantageous to select initially for a marginal change of phenotype in order to enrich for the heterozygote (Sections 3.5.3 and 3.5.5). For recessive antimetabolite-resistant variants, however, the difference in resistance between heterozygote and wild-type is usually not sufficiently large to permit a high enrichment ratio in a single-step selection. In optimizing conditions it is very useful to have available a cell line of the phenotype to be isolated, or, failing this, one that closely approximates it. One can then perform reconstruction selections by adding a small number of these cells to the population on which the selection is to be performed, and exposing the mixture to selective conditions. If the recovery of the two cell types can be scored independently on the basis of colony morphology or by some other means, one can then estimate variant recovery and enrichment factor directly. If not, one can derive these parameters from a comparison of the total frequency of recovery for the mixture with that obtained with wild-type cells alone. This is less satisfactory as the two cell types may influence each other's behavior in the selective system (compare Section 2.1.7).

Use of different isolation techniques in succession may be desirable either if the first selection technique is not completely specific for the desired variant phenotype and incapable of enriching the population beyond a certain proportion, or where one wishes to avoid repeated exposure to a selective agent—if, for instance, it is mutagenic. An example of the second situation is to be found in the isolation of an amino acid transport variant by a single round of tritium suicide followed by replica plating (180).

3

CELL VARIANTS

Sources of heritable variation in cultured cells are of four principal kinds. First, interspecies difference can provide useful genetic markers such as the difference in sensitivity of rodent and primate cells to ouabain, the latter cells being sensitive to concentrations some four orders of magnitude lower than those required to kill rodent cells (241). Second, a source of variation is to be found in cell lines isolated from human patients with inborn errors of metabolism or from mutant strains of laboratory animals (see 242 for references). Third, some cell lines have extra nutritional requirements which cannot be accounted for by a germ-line lesion in their organism of origin (243–245). To what extent this reflects tissue-specific gene regulation and to what extent adaptation to the tissue-culture environment is not clear. Fourth, variant cells isolated by selection or screening processes in culture are the ones with which this chapter will be primarily concerned. Variants arise spontaneously in all cell lines: Some lines produce variants at an elevated frequency because of the presence of "mutator" genes (Section 3.1). The frequency of variant production may also be increased by the application of physical or chemical mutagens (Section 3.2) or, to a lesser extent, by the use of viruses (Section 3.3). A number of factors influence the decision whether such a mutagen should be used in any particular isolation procedure. If the frequency with which a class of variant is generated spon-taneously is very low, it may be impractical to isolate such a variant without the use of a mutagen to boost the frequency. In other instances it may be desirable to produce variants by a particular mechanism, for example, single base transition, frameshift, or large deletion, or to produce the widest possible spectrum of variants, and here the use of one or more mutagens that induce particular kinds of mutational event may be advantageous. On the other hand, the use of a mutagen increases the risk of second site mutational events at irrelevant loci which may cloud the interpretation of the effects of the lesion of interest: There exists at present no mechanism for the elimination of such unwanted mutations from cultured cells such as exists in microorganisms or intact mammals where one can backcross to the wild type.

By proper design of the isolation procedure it is possible to obtain a number of independently arising variants from a single wild-type clonal line. It is important to realize that variants are generated spontaneously throughout the lifetime of the line rather than at the moment at which selection is applied, so that in order to guarantee independent origin for spontaneous variants it is necessary to isolate (wild-type) subclones from the starting line and then derive a single variant from each. Alternatively, a physical or chemical mutagen may be applied for less than one cell generation time to a single culture, which is divided into subcultures immediately afterward. This ensures that mutagen-induced variants in different subcultures are independent of each other, but as there will probably also be spontaneous variants present the firmest conclusion that one can draw about independence of origin is the following: Provided that it can be demonstrated that mutagenesis appreciably increases the yield of variant cells, there is a high probability (which is, notwithstanding, less than unity) that variants isolated from different subcultures will be of independent origin. This conclusion may be sufficient in a study using a large number of variants.

3.1. HIGHLY MUTABLE CELL LINES

Certain alleles of a number of bacterial genes cause an increase in the spontaneous mutation frequency. These "mutator" genes include those for DNA polymerase and for elements of the excision repair system which repairs damage to DNA caused by UV radiation (246). Similar variants have recently been discovered in mammalian cell lines. A variant resistant to aphidicolin, an inhibitor of DNA polymerase α, shows an increase in the spontaneous frequency of mutation to ouabain resistance (247). A number of variants sensitive to ultraviolet radiation, falling into at least five complementation groups, are hypermutable by UV radiation at three loci (248–250). Increased susceptibility to induction of mutations by UV radiation is also seen in cells from xeroderma pigmentosum (XP) patients. Here the increase in mutability correlates with the extent of the deficiency in excision repair caused by the XP lesion, and is observed with certain chemical mutagens as well as UV (251). A class of thymidine-requiring (THY$^-$) variants resistant to 5-azacytidine and 5-fluorouracil show an increase in mutation frequency both to ouabain resistance and to thioguanine resistance, but not at two other loci (252,253). Selection for resistance to cytotoxic drugs by "training" procedures (Section 2.1.1) may also enrich for variants with an increased spontaneous mutation rate (254).

3.2. PHYSICAL AND CHEMICAL MUTAGENS

Table 3.1 lists a number of physical and chemical mutagens that have been used successfully with cultured cells. In general, characterization of the

TABLE 3.1. SOME PHYSICAL AND CHEMICAL MUTAGENS USEFUL FOR THE INDUCTION OF MUTATIONS IN TISSUE-CULTURE CELLS

Mutagen	Predominant Type of Mutational Event	References to Successful Use in Cultured Cells
N-methyl-N'-nitro-N-nitrosoguanidine (MNNG)	Transition + misrepair (259,260)	92,[a] 260,[b] 265,[c] 266[d]
Ethyl methanesulfonate (EMS)	Transition + misrepair (256,267)	92, 255,[e] 260, 265, 266
Methyl methanesulfonate (MMS)	Misrepair (256)	260, 266
ICR-191	Frameshift (267)	255, 265
8-Methoxypsoralen	Frameshift (268,269)	257[b]
1,2,7,8-Diepoxyoctane (DEO)	Deletion (270,271)	272[b]
X- and γ-radiation	Misrepair (256)	273–275,[b] 276[b,f]
UV radiation	Misrepair (256)	273,[b] 277[b]

Assay systems:

[a] CHO cells; auxotrophy and proline prototrophy.

[b] Chinese hamster fibroblasts; HGPRT deficiency.

[c] L5178Y lymphoma; asparagine prototrophy.

[d] L5178Y lymphoma TK heterozygote; TK deficiency.

[e] Y5606 myeloma; immunoglobulin deficiency.

[f] Human diploid fibroblasts; HGPRT deficiency.

predominant type of mutational event induced by these agents has been carried out in microorganisms: Most assay systems used for cultured mammalian cells are useful only for the analysis of mutation frequency, although a recently described system involving the screening of colonies of the Y5606 myeloma for the presence of secreted immunoglobulin by a precipitation technique (255) holds promise for the characterization of mutagenic events by analysis of the gene product. In the meantime, extrapolation of conclusions about mechanisms drawn from microorganism studies should be made cautiously, particularly where the repair system of the cell may play an important role. Cell types may differ not only in the relative contribution made by misrepair mutagenesis, but in the relative frequency of different kinds of event generated by misrepair which may include transitions, transversions, frameshifts, and deletions (256).

An important variable in the design of a mutagenesis protocol is the expression time, that is, the time allowed to elapse between the end of the mutagen treatment and the application of selective conditions. For short expression times the yield of mutants increases as the expression time is lengthened: This may reflect the necessity for processes such as activation of the mutagen, conversion of the primary lesion to a viable genetic change by cellular repair mechanisms, or dilution of preexisting gene product (257). In early work in which the induction of azaguanine-resistant derivatives of Chinese hamster fibroblasts was used as an assay system it appeared that the yield of mutants reached a peak, beyond which it then fell as the expression time was further increased. However, this was due to the design of the selective system: Cells were treated with mutagen, plated at a fixed density, and the selective agent then added at various times thereafter (the so-called *in situ* technique). This led to higher cell density at later expression times, with consequent loss of mutant cells by metabolic cooperation with wild-type cells (Section 2.1.7). If cells are replated at a fixed density at the end of the expression time, or if a suspension cell line is used, the yield of mutants reaches a plateau and does not show a decrease when the expression time is further increased (257). Nevertheless, such a fall would be expected for any mutant with a slower growth rate than its parent, and it is clear that optimal expression times will vary with the class of mutant under study.

While some assay systems, such as those based on mutation to HGPRT deficiency, will detect any event leading to inactivation of the gene product, others are more specific. An interesting example of the latter group is found in the assay of mutation to ouabain resistance. Ouabain-resistant cells arise by an alteration in the plasma membrane Na^+,K^+-ATPase which inactivates its ouabain binding site while leaving its Na^+ ion pumping capacity intact (Section 3.5.3): It follows that a relatively subtle change in the protein is required and therefore while an increase in the frequency of ouabain-resistant cells is readily induced with agents that cause single-base transitions, no such increase is demonstrable with frameshift mutagens or with γ-radiation (257,258).

The most efficient mutagens currently available for use with cultured cells are MNNG and EMS. MNNG and, to a lesser extent, EMS cause mutational events preferentially at the replication fork of DNA (256,259). This can lead to the induction of multiple, closely linked mutations which may be undesirable. The mutagenicity of MNNG is enhanced by high intracellular thiol concentrations (256). As MNNG is rapidly broken down in thiol-containing media (260) it seems likely that the product of the reaction with thiol is an intermediate in the mutagenic process. However, until more is known about the stability of this intermediate and its ability to enter cells, it seems wiser to allow the reactions to occur intracellularly, rather than to add thiol to the medium used for the mutagenesis treatment.

Colcemid has been used in attempts to stimulate chromosome nondisjunction in cultured Chinese hamster cells (261). However, no stable clones with less than the parental diploid chromosome number were obtained, all the surviving clones being either tetraploid or intermediate between diploid and tetraploid. Nondisjunction of a specific chromosome, or a lesion in a chosen region of a chromosome, visible as a change in banding pattern, can be achieved by laser microirradiation (262). However, the high cost of the equipment necessary for this technique has so far restricted its application.

Special considerations apply to the design of mutagenesis protocols where one wishes to increase the yield of mutations in mitochondrial DNA (mtDNA), because of the large number of copies present per cell. The most successful regimes have involved depletion of the pool of mtDNAs prior to mutagenesis, and selection for the variant phenotype during repopulation of the cell from the surviving mitochondria (263). A number of depletion techniques are available, including ethidium bromide treatment, removal of cytoplasm by enucleation, selective labeling of mtNDA of TK$^-$ cells with BUdR followed by visible light irradiation, and treatment with rhodamine 6G (263,264). Two strategies have been employed for selective mutagenesis of mtDNA: One uses the above-mentioned selective incorporation of BUdR in TK$^-$ cells, while the other employs cycloheximide to arrest nuclear DNA replication, in combination with mutagens specifically or preferentially active on replicating DNA, such as 2-aminopurine or MNNG (263).

Finally, mention should be made of undesirable mutagenic effects to which cells may be exposed during routine culture. In particular, the action of fluorescent light on tissue cultures can produce chromosome abnormalities, possibly via the action of photoproducts of riboflavin (19).

3.3. VIRAL MUTAGENS

In view of the importance of transposable elements (278) as insertion mutagens in bacteria, transforming viruses that integrate into the host genome might be expected to have a mutagenic effect in mammalian cells. Infection with SV40 increases mutation frequencies in human and Chinese hamster cells but the effect appears to be associated with gross chromosomal aberrations and no effects due to insertion within specific genes have so far been demonstrated (279). Moloney murine leukemia virus has been shown to act as an insertion mutagen in Rous sarcoma virus-transformed B31 rat cells, but the increase in mutation frequency due to the virus was small compared with the background frequency (280). An insertion mutagen giving a substantial increase in mutation frequency and detectable by the use of a nucleic acid probe would be extremely valuable as it would make it possible to isolate any gene in which mutants are selectable.

3.4. SURVEY OF VARIANT CELL LINES

The number of variant cell lines isolated is now very large and a detailed discussion of the properties of all of them is beyond the scope of this book. A number of reviews have surveyed both cell variants in general and particular categories of variant (281–285) and were invaluable in preparing Tables 3.2–3.11, which attempt to categorize the variants isolated to date. The categorization is one of convenience and I recognize that many variants could be classed in more than one category. While no tabulation of this kind can claim to be exhaustive I have tried to be as comprehensive as possible.

Given the extensive nutritional requirements of mammalian cells, the auxotrophic variants listed in Table 3.2 represent a substantial proportion of those theoretically isolable. All behave recessively in fusion hybrids, allowing complementation analysis which has been pursued most fruitfully in the case of the adenine auxotrophs where nine complementation groups have been identified and the corresponding biochemical deficiencies assigned to individual steps in the purine biosynthetic pathway. The uridine auxotrophs are of particular interest in view of the light which they shed on the organization of the mammalian genome. Three complementation groups are sufficient to account for all six steps in the biosynthesis of UMP. Variants in the first group, UrdA, have lost the first three activities, that is, carbamyl phosphate synthetase, aspartate transcarbamylase, and dihydroorotase: Revertants coordinately regain all three activities. This behavior is due to the presence of all three activities in a single polypeptide (286). Variants in the UrdC complementation group suffer a coordinate loss in the last two

TABLE 3.2. AUXOTROPHIC VARIANTS

Requirement	Selective System	References
Glycine (4)[a]	BUdR[b] suicide	92, 293, 294
Proline	None	295
Adenine (9)[a]	BUdR suicide	293, 296–304
Glycine, adenine, thymidine	BUdR suicide	296, 297, 305–307
Glycine, hypoxanthine, thymidine	^{3}H-deoxyuridine suicide	308
Adenine, thymidine	BUdR suicide	296, 297
Thymidine	(i) Resistance to araC[c] + 5FU[d]	
	(ii) ^{3}H-deoxyuridine suicide	252, 309
Uridine (3)[a]	(i) BUdR suicide	
	(ii) DAP[e] suicide	
	(iii) 5-FU resistance	
	(iv) 5-Fluoro-orotic acid resistance (in presence of uridine)	
	(v) Replica plating	310–316
Sterols	(i) BUdR suicide	
	(ii) Filipin resistance	
	(iii) Amphotericin B resistance	317–320
Serine	BUdR suicide	321
Glutamate	BUdR suicide	322
Alanine	BUdR suicide	322
Inositol	BUdR suicide	92
Asparagine	(i) FUdR[f] suicide	182, 323
	(ii) ^{3}H-thymidine suicide	
CO_2, asparagine	BUdR suicide	324, 325
Oleate	BUdR suicide	326
Myoinositol	Replica plating	238

[a] Number of complementation groups.
[b] 5-Bromodeoxyuridine.
[c] Cystosine arabinoside.
[d] 5-Fluorouracil.
[e] 2,6-Diaminopurine.
[f] 5-Fluorodeoxyuridine.

TABLE 3.3. TEMPERATURE-SENSITIVE VARIANTS

Altered Function	References
Aminoacyl tRNA synthetase (8)[a]	287, 288, 327–336
28s rRNA synthesis	337, 338
DNA synthesis	339–343
Hypoxanthine guanine phosphoribosyl transferase	344, 345
RNA polymerase II	346, 347
Thymidine kinase	348
Density-dependent growth inhibition and related properties	349–355
Glycine, hypoxanthine, and thymidine synthesis	356
Glycine, adenosine, and thymidine synthesis	307
Galactose utilization	357
Colchicine resistance	358
Adhesion	359
Others including cell cycle	360–370

[a] Number of complementation groups.

activities of the pathway, orotate phosphoribosyl transferase and orotidylate decarboxylase. Corresponding to these two complementation groups over-producer variants can also be isolated (Table 3.4) with coordinate increases in the same groups of activities.

The temperature-sensitive variants listed in Table 3.3 embrace both heat-sensitive and cold-sensitive phenotypes. By analogy with temperature-sensitive mutants in microorganisms it seems likely that these represent missense structural gene mutations, and in some cases it has been possible to demonstrate temperature-sensitive enzyme activity *in vitro* (287,288). As well as variants temperature-sensitive for cell growth it is possible to isolate variants with a temperature-sensitive lesion in a defined function for which both forward and reverse selective systems are available (see Table 3.12), by forward selection at the nonpermissive temperature followed by reverse selection at the permissive temperature. In microorganisms such temperature-sensitive variants are often most conveniently obtained as second-site revertants of an initial missense null mutant, and this approach may prove fruitful with mammalian cells also.

Table 3.4 lists variants with increased activities of specific proteins. In the cases of variants overproducing folate reductase and those overproducing

TABLE 3.4. VARIANTS WITH INCREASED ACTIVITIES OF SPECIFIC PROTEINS

Protein	Selective System	References
Dihydrofolate reductase	(i) Methotrexate resistance (ii) Fluorescein-methotrexate fluorescence	371
Ribonucleotide reductase	Resistance to hydroxyurea or deoxynucleosides	372, 373
Carbamyl phosphate synthetase, aspartate transcarbamylase + dihydroorotase	N(phosphonoacetyl)-L-aspartate resistance	291, 374–377
Orotate phosphoribosyl transferase + orotidylate decarboxylase	Resistance to pyrazofurin and 6-azauridine	378, 379
Asparagine synthetase	β-aspartyl hydroxamate resistance	380
Ornithine decarboxylase	Resistance to α-methyl ornithine and α-difluoromethyl ornithine	381
Adenylate deaminase	Resistance to coformycin-potentiated adenine toxicity	382
Component(s) of enzyme system catalyzing conversion of glutamate to glutamic semialdehyde	Azetidine 2-carboxylic acid resistance	383
Glutamine synthetase	Methionine sulfoximine resistance	384, 385
Argininosuccinate synthetase	Canavanine resistance	386
Metallothionein-I	Cd^{2+} resistance	387

TABLE 3.5. ANTIMETABOLITE-RESISTANT VARIANTS WITH ALTERED NUCLEIC ACID PRECURSOR METABOLISM

Antimetabolite	Affected Enzyme	References
(i) 8-Azaguanine (ii) 6-Thioguanine (iii) 8-Azahypoxanthine	Hypoxanthine guanine phosphoribosyl transferase	281, 388, 389
(i) 8-Aza-adenine (ii) 6-Mercaptopurine (iii) 2,6-Diaminopurine (iv) 2-Fluoroadenine	Adenine phosphoribosyl transferase	281, 390–392
(i) Adenosine (ii) Toyocamycin (iii) Tubercidin (iv) 6-Methylthiopurine riboside (v) 2-Fluoroadenosine	Adenosine kinase	281, 393–396
Adenine arabinoside	(i) Adenosine kinase (ii) Ribonucleotide reductase	397 397
Deoxyadenosine	Deoxyadenosine kinase	398
6-Thioguanosine	Purine nucleotide phosphorylase	399, 400
(i) 5-Bromodeoxyuridine (ii) 5-Fluorodeoxyuridine (iii) 5-Iododeoxyuridine (iv) Trifluorothymidine	Thymidine kinase	281, 401–404
5-Bromodeoxycytidine	Deoxycytidine deaminase	405
Cytosine arabinoside	(i) Deoxycytidine kinase (ii) Ribonucleotide reductase	406 406
(i) ³H-Uridine plus fluorouridine (ii) 6-Azauridine	Uridine kinase	407–409
Aphidicolin	Ribonucleotide reductase	410
(p-Hydroxyphenylazo) uracil	Ribonucleotide reductase	411

TABLE 3.6. ANTIMETABOLITE-RESISTANT VARIANTS WITH
ALTERATIONS IN NUCLEIC ACID OR PROTEIN SYNTHESIS

Antimetabolite	Site of Lesion	References
Aphidicolin	DNA polymerase	247
α-Amanitin	RNA polymerase II	346, 412–417
(i) Emetine (3)[a]	40s ribosomal subunit	418–423
(ii) Cryptopleurine		
(iii) Tylocrebrine		
Trichodermin	60s ribosomal subunit	424
Diptheria toxin	Elongation factor EF2	425–428

[a] No. of complementation groups.

the first three activities of the pyrimidine biosynthetic pathway it has been demonstrated that the increase in activity is due to gene amplification. The amplified unit extends beyond the gene itself and includes flanking sequences that vary in length. The other variants are as yet less well characterized but almost certainly include other instances of gene amplification. Recent, largely unpublished, work in this area is reviewed in ref. 289. It seems likely that the frequency with which one encounters such instances in mammalian cells compared with their relative scarcity in bacteria is related to the existence in mammalian cells of repetitive DNA (290): We may hope that the study of the genesis of overproducer variants will elucidate the mechanisms by which multigene families have evolved. It has recently been demonstrated that the amplification frequency depends on the chromosomal position of the gene, suggesting that flanking sequences play an active role in the amplification process (291).

Other variants resistant to toxic antimetabolites are grouped in Tables 3.5–3.8 according to the site of their lesion. As discussed in Section 2.1.1, resistance to a given antimetabolite may arise by more than one mechanism, and this accounts for appearance of the same antimetabolite in more than one table, for example, methotrexate in Tables 3.4, 3.7, and 3.8. The variants listed in Table 3.5 have been extensively exploited in the isolation of fusion hybrids and transfectants: Since the forward mutation inactivates a salvage pathway for utilization of exogenous purines or pyrimidines, thereby requiring the cell to use its *de novo* biosynthetic pathways, reverse selection can be achieved by the use of antimetabolites which block the latter pathways (Table 3.12). Also worthy of particular mention are tumor

TABLE 3.7. ANTIMETABOLITE-RESISTANT VARIANTS WITH ALTERED
CELL SURFACE COMPONENTS

Antimetabolite	Site of Lesion	References
Lectins	Membrane glycoproteins and glycolipids	285, 429, 430
Tunicamycin	Membrane glycolipids?	431
Anti-HLA + complement	HLA antigen	190
Glucocorticoids	Steroid receptors	432–435
Ouabain	Na^+,K^+-ATPase	436
Methotrexate	Transport	437, 438
5-Fluorotryptophan	Transport	439, 440
L-Phenylalanine	Transport	441
^{3}H-Amino acid suicide	Transport	180, 442
Chromate	Sulfate transport	443
Methylglyoxal bis-guanyl hydrazone	Polyamine transport	444
Canavanine	Transport	445
Colchicine	Membrane glycoprotein	446–448
2-Deoxy[1-^{3}H]glucose	Transport	449
^{3}H-Fucose	Cell surface glycoproteins	450
[2-^{3}H]Mannose	Cell surface glycoproteins	451
Methionine	Transport	452
LDL[a] reconstituted with 25-hydroxycholesteryl oleate	LDL receptor-mediated endocytosis	453
VM26,[b] VP16-213[b]	Membrane permeability	454

[a] Low-density lipoprotein.

[b] Anticancer drugs, derivatives of podophyllotoxin.

cell variants resistant to the lectin wheat germ agglutinin which exhibit
altered metastatic properties (292) and variants with altered mitochondria
(Table 3.8). In many, but not all, of the latter, the altered phenotype is
determined by a mitochondrial gene and is therefore cytoplasmically inherited,
a fact that is put to use in the selection and analysis of "cybrids" (Section
4.5).

TABLE 3.8. MISCELLANEOUS ANTIMETABOLITE-RESISTANT VARIANTS

Antimetabolite	Site of Lesion	References
(i) Colcemid	β-Tubulin	455–457
(ii) Colchicine		
(iii) Griseofulvin		
Taxol (resistant)	α-Tubulin	458
(resistant + dependent)	Spindle assembly	459, 460
(i) Oligomycin	Mitochondrial oxidative	461–464
(ii) Rutamycin	phosphorylation	
(iii) Venturicidin		
Antimycin	Mitochondrial electron transport	465
Chloramphenicol	Mitochondrial protein synthesis	263, 466, 467
Erythromycin	Mitochondrial protein synthesis	468
Sindbis virus	Translational control?	469
2-Deoxyglucose	Galactokinase	470
Methotrexate	Dihydrofolate reductase (structural)	437, 471–473
Pactamycin	Morphology	474
Puromycin	Not known	281
Adenine arabinoside	S-Adenosyl homocysteine hydrolase	397
L-Azetidine 2-carboxylic acid	Feedback inhibition of glutamic semialdehyde synthesis	475
25-Hydroxycholesterol	Feedback inhibition of hydroxymethyl glutaryl CoA reductase	476
Actinomycin D	Not known	281
Podophyllotoxin	Not known	477
Diphtheria toxin (cross-resistant to RNA viruses)	Endosome acidification	478, 479
Methylmercaptopurine riboside + adenine + uridine	Feedback inhibition of phosphoribosyl pyrophosphate synthetase	480

TABLE 3.9.   MISCELLANEOUS VARIANTS ISOLATED BY OTHER TECHNIQUES

Site of Lesion	Selective System	References
Acetylation of glucosamine 6-phosphate	Detachment from substratum in presence of PGE_1[a] and MIX[b]	481, 482
Adhesion	Lack of attachment to collagen substratum	483
Metabolic cooperation	"Kiss of death" (Section 2.1.7)	484–486
	Coculture with ouabain-resistant cells in ouabain and retinoic acid	487
H2-K^k antigen (structural)	FACS[c] (see Section 2.2.3)	218
Branched chain amino acid transaminase	BUdR[d] suicide in medium containing α-ketoisovalerate in place of valine	488
Induction of (2'-5') oligoadenylate synthetase by interferon	Screening for resistance to antiviral effect of interferon	489

[a] Prostaglandin E_1.

[b] 1-Methyl 3-isobutyl xanthine, an inhibitor of cyclic AMP breakdown.

[c] Fluorescence-activated cell sorting.

[d] 5-Bromodeoxyuridine.

TABLE 3.10.   VARIANTS ISOLATED BY NONSELECTIVE ISOLATION TECHNIQUES

Site of Lesion	References
Glucose 6-phosphate dehydrogenase	230
Adenine phosphoribosyl transferase (heterozygote)	231
Dihydroorotate dehydrogenase	315
Phosphatidyl choline synthesis	490
UV sensitivity (6)[a]	221, 237, 248
Mitomycin sensitivity	221
Ethyl methanesulfonate sensitivity	221
Ethanolamine phosphotransferase	491
α-Mannosidase	492
Glycoprotein synthesis	450
Alkaline phosphatase	493

[a] Number of complementation groups.

TABLE 3.11.   VARIANTS WITH ALTERATIONS IN DIFFERENTIATED FUNCTION[a]

Parent Cell Line	Isolation Technique	Altered Function	References
L6,L8 Rat myoblast	Passage after incubation at confluence	Differentiation into myotubes	494
L6,L8 Rat myoblast	Visual screening	Differentiation into myotubes	495
G3 Rat pituitary tumor	Passage and cloning	Prolactin synthesis	136
G3 Rat pituitary tumor	Passage and cloning	Growth hormone synthesis	136
MDCK (Canine kidney)	Tumorigenesis	Growth requirement for PGE₁[b] and insulin	119
MDCK (Canine kidney)	Growth in PGE_1-free medium	Growth requirement for PGE_1	119
Friend erythroleukemia	Cloning	High spontaneous level of differentiation	132
Friend erythroleukemia	Growth in presence of inducers of differentiation	Inducibility of differentiation	132
MOPC21 Myeloma	Screen clones by isoelectric focusing of secreted immunoglobulin	Immunoglobulin (structural)	496
Myeloma Hybridoma	FACS[c]	Heavy chain class	216, 217
Myelomas	Antibody precipitation screening	Ig secretion	497
Myelomas	Screen clones by isoelectric focusing of secreted products	Ig secretion	498
Lymphoma	Complement-mediated cytotoxicity	Membrane IgM	189
S49 Lymphoma	Not stated	β-Adrenergic receptor	499

Cell line	Selection method	Property	Reference
S49 Lymphoma	Terbutaline + inhibitors of cyclic nucleotide phosphodiesterase	Interaction between hormone binding and adenylate cyclase activation	499, 500
S49 Lymphoma	Isoproterenol + inhibitor of cyclic nucleotide phosphodiesterase	Adenylate cyclase	499, 501
S49 Lymphoma	Dibutyryl cAMP ± theophylline	cAMP-dependent protein kinase	499, 502–504
S49 Lymphoma	Dibutyryl cAMP + cytosine arabinoside	"Deathless"	359, 499
Molt-4 T-cell leukemia	FACS	Level of expression of antigen HTA-1	505
A431 Epidermoid carcinoma	Resistance to inhibition of proliferation by EGF[d]	EGF receptor	506
H4IIEC3 Hepatoma and derivatives	Morphology	Expression of several luxury functions	507
RPMI1640 Hamster melanoma	Visual screening	Melanogenesis	508
S91 Mouse melanoma	Visual screening	Melanogenesis	509
S91 Mouse melanoma	MSH[e]	Resistance to growth inhibition by MSH	510

[a] For such variants isolated from embryonal carcinoma cells, see Table 6.1.
[b] Prostaglandin E₁.
[c] Fluorescence-activated cell sorting.
[d] Epidermal growth factor.
[e] Melanocyte stimulating hormone.

TABLE 3.12. BIDIRECTIONAL SELECTIVE SYSTEMS

Variant	Forward Selection	Reverse Selection	References
Auxotrophs	See Table 3.2	Medium deficient in required metabolite	See Table 3.2
Temperature sensitive	See Section 2.1.4	Non-permissive temperature	284
HGPRT[a]	(i) 6-Thioguanine (ii) 8-Azaguanine (iii) 8-Azahypoxanthine	HAT (hypoxanthine + aminopterin + thymidine)	280, 388, 389, 511
APRT[b]	(i) 8-Aza-adenine (ii) 6-Mercaptopurine (iii) 2,6-Diaminopurine (iv) 2-Fluoroadenine	(i) AAT (adenine + aminopterin + thymidine) (ii) Adenine + azaserine	280, 390–392
Adenosine kinase	(i) Toyocamycin	Adenosine + alanosine + uridine	280, 393–396

	(ii) Tubercidin		
	(iii) 6-Methylthiopurine riboside		
	(iv) 2-Fluoroadenosine		
	(v) Adenosine		
Thymidine kinase		HAT	280, 401–404, 511
	(i) 5-Bromodeoxyuridine		
	(ii) 5-Fluorodeoxyuridine		
	(iii) 5-Iododeoxyuridine		
	(iv) Trifluorothymidine		
Deoxycytidine deaminase	5-Bromodeoxycytidine	Hypoxanthine + aminopterin + 5-methyldeoxycytidine	405
Uridine kinase	(i) ³H-Uridine + fluorouridine	Adenosine + uridine	407–409
	(ii) 6-Azauridine		
Metabolic cooperation	"Kiss of death"	"Kiss of life"	203

[a] Hypoxanthine guanine phosphoribosyl transferase.
[b] Adenine phosphoribosyl transferase.

3.5.   DO VARIANT CELL LINES REPRESENT TRUE MUTANTS?

Early studies on the isolation of variant cell lines revealed some properties not shared by mutant microorganisms which raised the question of whether they did indeed carry true mutations in the sense of alterations in DNA sequence, or whether their altered properties were due to so-called epigenetic changes, that is, changes such as those which occur in differentiation and switch the cell to an alternative heritable pattern of gene expression without a necessary change in DNA sequence. One observation raising such questions was that it was possible to "train" resistant variants by exposure to progressively increasing concentrations of antimetabolites (see Section 2.1.1), and that these variants were often unstable in the absence of continued selective pressure. Another was the relative ease with which it was possible to obtain, from near-diploid cell lines, variants that behaved recessively in fusion hybrids and that could be shown not to be X-linked, an observation difficult to reconcile with the necessity to mutate both alleles of an autosomal gene. Furthermore, the frequency with which variants could be obtained was reported to be independent of ploidy (512,513). It was because of these doubts that the term "variant" was introduced to avoid prejudging the mechanistic basis of the altered phenotype.

The only truly unambiguous criterion for classifying a variant as a true mutant is a change in DNA sequence, or, in the case of a structural gene mutant, a change in the primary sequence of the protein product. Needless to say, this has been demonstrated for very few somatic cell variants (or, for that matter, for few mutants in microorganisms) and a number of secondary criteria have become established which permit variants to be designated probable mutants. However, as none are unambiguous it is important to be clear about the conclusions that can be drawn if a variant cell line satisfies a particular criterion, and I shall therefore discuss them in some detail.

1.   The phenotype should be stable for multiple generations in all progeny cells (excluding, of course, rare revertants) in the absence of selective pressure. This criterion is necessary but clearly not sufficient.

2.   The events that give rise to the altered phenotype should occur with a random time distribution throughout the growth of the population prior to selection, and not in response to the application of the selective conditions. This may be tested by the Fluctuation Test originally devised for bacteria

by Luria and Delbruck (514). In this test, a series of cultures is established from a population of wild-type cells, the number of cells used to initiate each culture being so small that the probability of a variant cell being present at the start of the culture can be ignored. These cultures are then grown up into large populations and exposed to selective conditions in order to determine the number of variant cells present in each. If variants are generated throughout the growth period, then by chance in some cultures a variant will arise early, while the population size is still small, and therefore its progeny will account for a large proportion of the total cells present, whereas in others where variants do not arise until later they will account for a smaller proportion. The proportions for the different cultures will therefore show a much larger variance than would be obtained with replicate determinations on the original population. If variants are generated in response to the selective conditions, then these two variances should be identical. It is important to be clear that the Fluctuation Test gives information about the time distribution of the events giving rise to the altered phenotype and not about their mechanism, so that this criterion also is necessary but not sufficient to classify a variant as a mutant. Furthermore, Fluctuation Tests are less satisfactory for mammalian cells, particularly those that grow in monolayer, than for bacteria. Not only is the efficiency of recovery of variant cells under selective conditions usually much less than 100% but also it can be influenced by a number of factors that can introduce additional variability between cultures, such as the effects of dispersing agents on viability and the effects of cell interactions on the selective system. The problem of efficiency of recovery can be overcome by including reconstructions with known mixtures of variant and wild-type cells, but the other problems are hard to eliminate.

3.   It should be possible by applying mutagens to the wild-type population to increase the incidence of cells showing the altered phenotype, and similarly by applying them to the variant population to increase the incidence of revertants. This criterion is neither necessary nor sufficient to establish a variant as a mutant. As discussed in Section 3.2, different categories of mutant show differing responses to different categories of mutagen. Furthermore, one must bear in mind that variant clones with a given phenotype may differ one from another in the underlying mechanism responsible for that phenotype, particularly if some have arisen spontaneously and others in response to mutagen treatment. Therefore, it is not valid to draw conclusions about the mechanistic basis of a spontaneous variant from the observation

that the incidence of variants of similar phenotype is stimulated by exposure to a mutagen.

4. It should be possible, where the variant behaves recessively in fusion hybrids, to map the complementing gene(s) to a particular chromosome or chromosome segment. This is a necessary but not sufficient condition for a recessive mutation. One should be clear that what one is mapping is not the lesion, but a chromosome region whose product is absent or inactive in the variant. It does not necessarily follow that the reason for its absence or inactivity is a mutation.

5. In the case of a structural gene mutant it may be possible to demonstrate the existence of an altered protein product or to isolate a revertant in which one can demonstrate this. This is clearly not a necessary condition: whether it is sufficient depends upon the technique used to demonstrate that the product is altered. One must bear in mind that the kinetic properties of an enzyme in a crude extract may be modified by association with other molecules, and that changes in electrophoretic mobility or even in peptide maps can be the result of changes in posttranslational modifications such as glycosylation as well as in primary sequence. Nevertheless, the demonstration of an altered protein product can be taken as a strong suggestion that one is dealing with a structural gene mutant.

In discussing the incidence of mutants in a population it is important to distinguish between the terms "mutation frequency" and "mutation rate." Mutation frequency is simply the proportion of mutant cells in a population at any given time and will vary from population to population. Mutation rate is the rate of production of mutants per cell per generation: It cannot be determined from a single mutation frequency and is obtained from an analysis of the results of a Fluctuation Test (see above). Strictly speaking, where the mechanistic basis of the lesion is unknown one should use the corresponding terms "variation frequency" and "variation rate" but unfortunately these have not gained wide usage.

In the following sections I discuss the nature of a few variants that illustrate the major categories found. I first consider HGPRT and myeloma immunoglobulin chain variants as examples free from the complications of multiple alleles. Then, after discussing the autosomal dominant ouabain- and methotrexate-resistant variants, I review what is known about the genesis of autosomal recessive variants. Finally, I consider chloramphenicol-resistant variants to exemplify extrachromosomal inheritance.

3.5.1. Hypoxanthine Guanine Phosphoribosyl Transferase (HGPRT) Variants

The *hgprt* gene is X-linked and therefore functionally hemizygous in all diploid mammalian cells. This, coupled with the availability of efficient forward and reverse selective systems, makes it particularly readily exploitable for genetic analysis and a large number of HGPRT⁻ variants have been isolated from a range of cell lines. Such variants arise at frequencies of around 10^{-6} in unmutagenized populations, and at higher frequencies after treatment with a variety of mutagens (Section 3.2). Some variants have no detectable HGPRT activity while residual activity is demonstrable in others (389). The former category may be subdivided into those which contain material cross-reacting with antibodies raised against wild-type enzyme (CRM⁺) and those which do not (CRM⁻). The cross-reacting material in CRM⁺ extracts may be analyzed by polyacrylamide gel electrophoresis: While in some cases its mobility is identical to that of wild-type enzyme, in others its mobility is changed, and in the latter cases differences have been found in tryptic maps, a single peptide being affected in each case (389). In some variants with residual activity it has been possible to demonstrate altered kinetic parameters, increased thermolability, and/or altered electrophoretic mobility. A particularly interesting example, seen in several variants, is an increase in K_m for phosphoribosyl pyrophosphate (PRPP). This reduces the rate of incorporation of hypoxanthine below detectable levels unless aminopterin is added to the medium, which results in an increase in intracellular PRPP concentration, so that the variants will grow both in azaguanine and in HAT (389).

The majority of HGPRT⁻ variants undergo reversion at low but detectable frequencies which can be increased by the use of mutagens. Some of the revertants have enzyme with altered kinetic properties or altered electrophoretic mobility compared with the wild-type enzyme. One CRM⁺ variant with a single peptide missing (or unresolved) in its tryptic peptide map gave rise to a revertant with a new peptide differing in mobility from that of the wild-type (389). Another variant gave a revertant with quantitatively altered immunoprecipitation properties. These changes are all consistent with reversion by second-site intragenic suppression. A further category of revertants has also been obtained with elevated levels of a variant enzyme protein of specific activity much less than that of the wild type (389).

Thus subject to the caveats stressed above about the interpretation of observations of altered gene product, there seems little doubt that HGPRT⁻ variants represent structural gene mutants. In one case (515) definitive evidence has been obtained which allows a variant to be identified as a nonsense mutant, namely, suppressibility by exogenous ochre-suppressor transfer RNA introduced by microinjection.

3.5.2. Immunoglobulin Chain Variants

Although immunoglobulin chains are the product of autosomal genes a phenomenon termed allelic exclusion (516) ensures that only one allele is expressed in the normal B lymphocyte lineage and in the corresponding tumors, the myelomas. Myeloma variants secreting structurally altered immunoglobulins are readily isolable by rapid screening techniques (Table 3.11). In four such spontaneous variants the difference was localized to the H chain, and primary sequence analysis identified the four changes as a nonsense mutation, an internal deletion of 282 bases, a frameshift mutation, and a missense mutation, respectively (517). However, one should be cautious in extrapolating conclusions from this system to somatic cell variants in general, because the mutation rates are unusually high and may be related to mechanisms for the generation of diversity in the immune response.

3.5.3. Ouabain-Resistant Variants

The steroid compound ouabain inhibits plasma membrane Na^+,K^+-ATPase, the so-called "sodium pump," which utilizes free energy obtained from ATP hydrolysis to pump Na^+ out of the cell, exchanging it with K^+, and thereby maintaining the cytoplasm at an ionic composition different from that of extracellular fluid or growth medium (518). Because of this inhibition ouabain is toxic to cultured cells although there is wide variation in the level of sensitivity between cells of different species, primate cells being some four orders of magnitude more sensitive than rodent cells (241). Somatic cell variants resistant to ouabain can be selected from sensitive cell lines: These variants possess an Na^+,K^+-ATPase with reduced sensitivity to ouabain (436). The frequency of such variants is increased by transition mutagens but not by frameshift mutagens or agents leading to gross changes in DNA, in keeping with the requirement to maintain a functional, though subtly modified, gene product (Section 3.2). Fusion hybrids between resistant and sensitive CHO lines exhibit an intermediate level of sensitivity, although for practical purposes, if the ouabain concentration is appropriately chosen, the resistance can be regarded as dominant. This is particularly true in rodent cells, where in medium of normal ionic concentration resistant × sensitive hybrids are resistant to ouabain at 3 mM, the limit of solubility (436). This dominant behavior readily explains the ease of isolation of ouabain-resistant variants from near-diploid cells, and indeed Lever and Seegmiller (258, 519) and Choy and Littlefield (520) have obtained evidence in support of the contention that selected variants are heterozygous in general. The latter authors obtained the more clear-cut results: They selected from diploid human fibroblasts a subline resistant to $10^{-6}\,M$ ouabain (an increase in resistance of about 100-fold compared with the wild type) and thence another subline resistant to $10^{-5}\,M$ ouabain. Cells of the first subline possessed 39% of the level of high-affinity ouabain receptors present in the wild-type line. Their uptake of $^{86}Rb^+$, an analogue of K^+, showed a biphasic response to variation in ouabain concentration, with a resistant phase accounting for about 40% of the total. In contrast, the subline resistant to $10^{-5}\,M$ ouabain possessed only 8% of the wild-type level of high-affinity receptors and its $^{86}Rb^+$ uptake was uniformly ouabain-resistant, while that of the wild-type was uniformly ouabain-sensitive. This strongly suggests that the first step of their selection, corresponding to the one-step selections normally employed, yielded the heterozygote and the second step the homozygote. The resistant phase of $^{86}Rb^+$ transport in the heterozygote is indeed predicted to be less than 50% of the total, since binding of ouabain to either of two identical subunits of the trimeric Na^+,K^+-ATPase is sufficient to inhibit transport (see discussion in ref. 520).

3.5.4. Methotrexate-Resistant Variants

Methotrexate, the 4-amino analogue of folate, inhibits the enzyme dihydrofolate reductase (DHFR). Variants resistant to methotrexate may arise by three mechanisms: reduced ability to transport methotrexate (Table 3.7), altered affinity of DHFR for methotrexate (Table 3.8), and increased levels of DHFR due to amplification of the structural gene (Table 3.4). Variants of the third class have been isolated by "training" procedures (Section 2.1.1). The resistance may be stable or unstable, the former being typical of Chinese hamster variants and the latter of mouse variants. In all cases of stable amplification the genes are chromosomal and in general can be localized to an expanded region of a single chromosome which shows no substructure in trypsin-Giemsa banded metaphase spreads and is therefore denoted a homogeneously staining region (HSR). In Chinese hamster cell lines the HSR is present on the long arm of one homologue of chromosome 2, in the region to which the nonamplified *dhfr* gene has been mapped (371). In contrast, in unstably amplified cell lines, where around half the amplified genes are lost within 20 cell doublings in nonselective medium, the *dhfr* genes are present on extrachromosomal elements called double minute chromosomes (DMs) which are self-replicating but do not contain centromeric staining regions and may be lost by micronucleation or unequal segregation at mitosis (371). Where a structural gene mutation has occurred after amplification it is possible to follow changes in the proportion of the different alleles within one cell in response to selective pressure (371), a process that may well be of wider evolutionary significance (521). Methotrexate resistance when due to stable gene amplification is dominant in fusion hybrids and transfectants. The incidence of this class of variant is unaffected by mutagens.

3.5.5.   Autosomal Recessive Variants

As discussed above, two observations that raised questions about the mechanistic basis of cell variants were the relative ease with which autosomal recessive variants could be obtained from near-diploid cells and the effects of ploidy on variant frequency. The original experiments on the basis of which it was concluded that ploidy had no effect on variant frequency (512,513) were subsequently criticized on detailed technical grounds (522). More careful measurements have revealed a difference in the frequency of induction by EMS of HGPRT$^-$ variants between diploid and tetraploid lines, although this was only 25-fold rather than the factor of more than 10^4 predicted on the basis of two independent mutational events (523). This discrepancy was explained by the subsequent observation that if tetraploid cells heterozygous for a null allele of another X-linked enzyme, glucose 6-phosphate dehydrogenase (G6PD), were selected for thioguanine resistance, about half the resistant clones had become G6PD$^-$ (524), as predicted if the events leading to inactivity of the two alleles of HGPRT were not identical, one being segregation of an X chromosome or part thereof. Such an event would not be tolerable for both X chromosomes because it would completely eliminate essential genes.

It appears that similar events may explain the generation of autosomal recessive variants in diploid cells. The frequencies with which recessive emetine-resistant variants are obtained in the CHO cell line is about 100-fold higher than in five other Chinese hamster cell lines, whereas the frequencies of ouabain- and thioguanine-resistant variants are similar (525). One of the other cell lines, V79, was subcloned and it was found that about 1% of clones reproducibly yielded emetine-resistant variants at frequencies of the same order as in CHO cells (526). This supports the suggestion that CHO cells and about 1% of the original population of V79 cells are functionally hemizygous at the emetine locus due to loss or inactivation of a chromosome segment (527). This suggestion is further strengthened by measurements of the frequency of production of emetine-resistant segregants from hybrids between emetine-resistant CHO variant cells and a series of Chinese hamster lines: Hybrids with wild-type CHO cells yielded segregants at a much higher frequency than hybrids with other Chinese hamster lines (528). The lesion responsible for functional hemizygosity at the emetine locus of CHO cells has now been identified as a cytologically visible deletion in one homologue of chromosome 2 (529).

Functional hemizygosity in CHO cells is not limited to the emetine locus. CHO variants resistant to α-amanitin contain only resistant RNA polymerase II while similar variants isolated from other cell lines contain both sensitive and resistant enzyme (417). Since this marker is dominant there should be no selective pressure favoring loss of the sensitive allele. Gupta (530) inserted into CHO cells a number of recessive resistance markers presumed to lie in functionally hemizygous parts of the genome because of their ease of isolation. He then hybridized the multiply marked cells with wild-type CHO cells and examined the segregation of resistant derivatives from the hybrid cells. None of the markers cosegregated, indicating that in CHO cells functionally hemizygous regions are scattered over the genome rather than being clustered on a single chromosome. However, CHO cells are not functionally hemizygous at the majority of loci: Electrophoretic shift variants isolated by nonselective means in general show the presence of both alleles (531). For most loci it therefore appears likely that functional hemizygosity is present only in a subpopulation of cells, as is the case with the emetine locus of V79 cells (see previous paragraph).

In other instances there is evidence that the derivation of autosomal-recessive variants is a two-step process, of which one step may be a segregation event occurring at relatively high frequency. Eves and Farber (532) isolated adenosine kinase-defective variants from a mouse cell line heterozygous for an electrophoretic variant of the esterase Es-10 which, like adenosine kinase, is coded by a gene on chromosome 14. Nine of twenty AK^- variants expressed only one Es-10 allele and were missing most or all of one copy of chromosome 14. In the remainder the two homologues of chromosome 14 were commonly found as an isochromosome. In the cases of thymidine kinase, adenine phosphoribosyl transferase, and glutamyl tRNA synthetase, revertants obtained by back-selection from defective variants have a much higher frequency of mutation to the defective state than does the wild type (266,336,390,533), suggesting that they represent the heterozygote. In the case of APRT, the presumed heterozygote can also be isolated directly from the wild type either by replica-plating or by selecting for marginal resistance (231,391). Analysis of *aprt* heterozygotes in CHO cells suggests that only one allele can undergo deletion rather than point mutation: This may be a position effect as the two alleles are carried on different CHO marker chromosomes (534,535).

3.5.6. Chloramphenicol-Resistant Variants

Chloramphenicol inhibits mitochondrial but not cytoplasmic protein synthesis, leading to growth inhibition of mammalian cells after several generations of exposure to the drug. A large number of chloramphenicol-resistant (CAP^R) cell lines have been isolated (263). While selection regimes have differed, "training" procedures appear to be advantageous in allowing the maximum opportunity for repopulation of the cell by mutant mitochondrial DNAs. The related compound Tevenel may have advantages over chloramphenicol itself in selection procedures since it has less secondary toxicity by inhibition of mitochondrial NADH oxidase (263). Cytoplasmic inheritance of chloramphenicol resistance has been convincingly demonstrated by a number of criteria. First, resistance can be transferred by fusion of enucleated fragments of resistant cells to sensitive cells (see Section 4.5). Second, during mitotic replication of hybrids between resistant and sensitive (CAP^S) cells, segregation of the resistant phenotype commonly occurs rapidly and without detectable chromosome loss, but in coordination with segregation of the corresponding parental mitochondrial DNA, which can be monitored by restriction endonuclease polymorphisms. Third, treatment of CAP^R cells with Rhodamine 6G prior to hybridization abolishes the ability to transfer resistance. Fourth, resistance can be transferred by purified mitochondrial DNA in certain cases. Definitive proof that the phenotype is due to a mutation in mitochondrial DNA has been obtained for three human and two mouse variants. In all cases, DNA sequencing localizes the mutation to one of two highly conserved regions near to the 3' end of the gene for the large mitochondrial rRNA. Mutations in the same regions are found in chloramphenicol-resistant strains of yeast (263). It has also been shown that the dominance of chloramphenicol resistance is not simply due to the independent expression of separate resistant and sensitive mitochondria within a cell, but that cooperation occurs between mitochondria such that the genes of both can be expressed in the presence of chloramphenicol. The mechanism of this cooperation is unknown: It may involve fusion of mitochondria or exchange of gene products between them. Whatever the mechanism, it seems likely that it is responsible for the observed persistence of mtDNA from the sensitive parent after many generations of selection in chloramphenicol (263).

3.5.7. Conclusions

The anomalous behavior which cast doubt on the possibility that somatic cell variants could be true mutants may now be regarded as largely resolved. Functional hemizygosity and segregation-type events account for the fact that it has been possible to isolate autosomal recessive variants and for the slightness of the effects of ploidy. The isolation of unstably resistant variants by "training" procedures can be explained in terms of gene amplification. It is, of course, a matter of semantics whether one considers gene amplification variants to be mutants or not. Well-studied variants such as those at the *hgprt* locus are almost certainly structural gene mutants, and definitive evidence is available that chloramphenicol-resistant variants arise by mutations in mitochondrial DNA. However, the mechanistic basis of the lesions in the vast majority of variants remains unknown, and now that the term "variant" has become established it seems reasonable to retain it except in those cases where mutant status has been unambiguously demonstrated.

4

CELL HETEROKARYONS AND HYBRIDS

Cell fusion occurs spontaneously at a low frequency when cells of different lines are co-cultured (99,536). Its frequency may be increased by exposure to a variety of agents (Section 4.1). The initial product of fusion, the heterokaryon, contains two or more distinct nuclei within a common cytoplasm (Section 4.2). A small minority of heterokaryons yield viable hybrids in which both genomes are combined within a single nucleus (Section 4.3). A cell fragment or organelle may serve as one fusion partner in place of an intact cell (Sections 4.5 and 4.6). For convenience chromosome-mediated gene transfer experiments are discussed in this chapter because the analysis to which they can be subjected parallels that for fusion experiments with X-irradiated cells (Section 4.7). DNA-mediated gene transfer will be discussed in Chapter 5.

4.1. FUSOGENS

In early work UV-inactivated Sendai virus was employed as a fusogenic agent (537) but has now been largely superseded by polyethylene glycol (PEG) which is simpler to use and has a wider spectrum of application. Since the initial description of the use of PEG as a fusogen (538) the protocol has been refined to circumvent problems of toxicity and to increase its efficiency. PEG is an efficient fusogen only within a narrow concentration range near 50% w/w where cytotoxic effects become significant (539). The toxicity may be reduced by omitting Ca^{2+} ions during the fusion and immediately afterward (539). Fusion is more efficient when the PEG solution is made up in 0.15 M HEPES buffer, pH 7.55, than in serum-free growth medium (540). pH is an important variable, and sterilization of PEG solutions by membrane filtration is preferable to the commonly used procedure of autoclaving the PEG which increases its acidity (540). The molecular weight of the PEG appears to be a variable that requires independent optimization for different cell types, since both fusogenic capacity and toxicity increase as the molecular weight is reduced from 6000 to 600, but reduction of the molecular weight to 200 eliminates the fusogenic capacity (539,540). A number of compounds act synergistically with PEG, allowing efficient fusion to be reached at lower PEG concentrations, and thus further reducing toxicity. The most commonly used of these is DMSO (541), but a number of other compounds are even more effective (542): Many are also effective cryoprotective agents (Section 1.10) and inducers of differentiation in erythroleukemia cells (178). PEG is a more efficient fusogen with cell monolayers than with suspensions of the same cell type, in spite of the fact that recent exposure to trypsin increases the efficiency of fusion (539). The use of an agglutinin such as PHA in combination with PEG enhances fusion in both configurations, and is particularly valuable for the fusion of nonadherent cells and subcellular fragments (539). Where only small numbers of suspension cells are available, recovery is aided by carrying out fusion on the surface of a cellulose acetate filter (543). The viability of adherent cell types after suspension fusion can be increased by the use of conditioned medium during the attachment period (539). A number of polymers other than PEG are also efficient fusogens, notably polyvinylpyrrolidone, polyvinyl alcohol, and polyglycerols (540).

A recent, highly promising development is electrofusion, that is, fusion as a result of the application of electric fields. Two such techniques have been described: In the first, suspension cells which have undergone clustering in response to an inhomogeneous alternating field ("mutual dielectrophoresis") are fused by the application of a single field pulse of high intensity and $2-50$-μsec duration (544), while in the second, monolayer cultures are fused by the application of repetitive square wave pulses of a few microseconds duration (545,546). Both methods yield heterokaryons efficiently, and for given cell types the extent of fusion can be regulated by varying the pulse amplitude and duration. Very high yields of viable hybrids have also been claimed for the technique (unpublished results cited in refs. 544 and 546) but as yet no details have been published.

4.2. HETEROKARYONS

Many of the kinds of genetic analysis possible with viable hybrids are possible in principle with heterokaryons, which appear to obey similar rules regarding coexpression or extinction of parental phenotypes (100; see, however, Section 4.3). In practice, however, the analysis is complicated by the presence of unfused parental cells and homokaryons in the population resulting from the fusion. Heterokaryons, being predominantly nondividing cells, cannot be isolated by the selective growth procedures applicable to viable hybrids. Techniques such as the use of irreversible biochemical inhibitors (Section 2.1.6) or physical separation by two-wavelength fluorescence-activated cell sorting (Section 2.2.3) after labeling the parental cells with different fluorochromes may allow this problem to be eliminated, but to date the use of heterokaryons has been largely restricted to the study of functions that can be analyzed at the single-cell level. These include the reactivation of dormant nuclei such as that of the erythrocyte and associated synthesis of DNA, RNA, cell surface antigens, and HGPRT, all of which can be monitored by autoradiography or other cytochemical techniques (99,100). The development of micromethods for biochemical analysis of small numbers of cultured cells isolated by microdissection (547) has made it possible to carry out complementation analysis on human inborn errors of metabolism using heterokaryons produced from diploid fibroblasts, including amniotic fluid-derived cells (548). Apart from problems of isolation, genetic analysis with heterokaryons suffers from the disadvantages that their lifespan is limited and that the interacting genomes are segregated in separate membrane-bounded nuclei. Counterbalancing these however are the advantages that genome dosage studies are possible at the single-cell level and that heterokaryons can be studied from the moment of fusion. The latter property is important both in allowing chromosome loss to be circumvented and in allowing analysis of time-dependent processes such as the cell cycle and differentiation. When mitotic cells are fused to interphase cells the interphase nuclei in the resulting heterokaryons undergo premature chromosome condensation (PCC): G_1 nuclei produce thin, extended filaments, while G_2 nuclei give rise to thicker, two-stranded filaments and S-phase nuclei to irregular, fragmented chromatin masses. This phenomenon has been put to many uses including the visualization of chromosomes in non-dividing cells, the study of the molecular architecture of interphase chromatin, the production of high-resolution chromosome banding patterns, and the detection of chromosome damage and monitoring of its repair (100,549). An example of the study of heterokaryons between cells at different points along a differentiation pathway is that of Wright (550), who observed induction of rat myosin light chain synthesis in heterokaryons between undifferentiated rat myoblasts and chick myotubes.

4.3. VIABLE HYBRID CELLS

The formation from heterokaryons of viable hybrid cells is a rare event (100) and therefore the use of selective methods for their isolation is almost obligatory. The best-known of such methods is the application of HAT selection to the fusion products of HGPRT⁻ and TK⁻ cells (511) but in principle any combination of counterselectable recessive markers (Table 3.12) and selectable dominant markers can be used provided that the two selective systems do not interfere. A case of such interference arises in the combination of HGPRT⁻ and APRT⁻ markers but this may be overcome by the use of a selective medium containing guanine, azaserine, adenine, and mycophenolic acid, the last-named compound inhibiting the conversion of IMP to XMP (551). If it is necessary to insert selective markers into cell lines for the sole purpose of hybridization the best strategy is to select sequentially for HGPRT deficiency and ouabain resistance in one parental cell line. This may then be fused to an unmarked cell line and hybrids selected in HAT + ouabain (552). This method has the advantages that both HGPRT⁻ and ouabain-resistant variants occur at acceptably high frequency and that a single marked line may be fused to a range of unmarked lines. For primate–rodent hybrids the interspecies difference in ouabain resistance may be used as a selective marker, so that selection of an HGPRT⁻ rodent parent is all that is required (241).

The chromosome complement of a hybrid cell is usually equal to the sum of the complements of one cell of each parental type, with some chromosome loss. Such cells are designated "1s + 1s," the symbol "s" denoting stemline chromosome number (Section 1.6). Some hybrid cells, however, contain more than one chromosome set of each parental type and these are designated "2s + 1s," and so on. Chromosome loss is usually slight in intraspecies hybrids but more extensive in interspecies hybrids, where it is common for the chromosomes of one species to be selectively lost (100; see also Section 4.4).

The study of hybrids enables one to examine the result of introducing genomes in different functional states into the same cell. These may involve the expression of "household" or "luxury" functions (Section 1.6).

4.3.1. Expression of Household Functions in Hybrid Cells

Unless one is dealing with variant cells a household function, by definition, is expressed by both parental cell types, and in all such cases expression occurs in the hybrid. Where the gene products of the parental cells can be distinguished, coexpression of the gene products of both parents can be demonstrated (100). This makes it possible to use such gene products as markers for the presence of chromosome segments in hybrids employed for mapping studies (Section 4.4). Where one fusion parent is a variant, the hybrid can be examined for "dominance" or "recessivity" of the variant phenotype: However, it should be noted that this is not an identical situation to that used in testing for dominance in classical genetics, since each unit of interaction may, and usually will, contain more than one allele. Where both parents are variants the hybrid can be examined for evidence of complementation (100). In all of the above situations the possibility of chromosome loss must be considered, so that absence of a function must be interpreted with care. For an intraspecies cross where chromosome loss is slight it may be sufficient to examine a number of independently arising hybrids, since this lessens the chance that they will all have lost the chromosome of interest. For interspecies crosses where chromosome loss is extensive this is not sufficient and it is essential to have cytogenetic or other evidence for the presence of the chromosome carrying the gene in question. Again, several independent hybrids carrying the chromosome should be studied in case some contain translocations.

4.3.2. Expression of Luxury Functions in Hybrids Between Cells of Different Histiotype

In general when a cell expressing a luxury function is fused to a cell of a different histiotype which does not express that function, the function is not expressed in the resulting hybrid, a phenomenon termed "extinction." However, if the ploidy of the expressing parent exceeds the ploidy of the nonexpressing parent there may be lack of extinction, sometimes associated with activation of the genome of the nonexpressing parent (reviewed in refs. 99,100, and 553). Extinction occurs at the heterokaryon stage and does not require nuclear fusion, but reexpression, followed in some cases by activation, may occur as hybrid colonies begin to proliferate, and it appears to be at this later stage rather than at the initial extinction step that the effect of gene dosage is exerted (554). Where reexpression occurs during this early phase of hybrid growth it may occur in the majority of hybrid cells, making it unlikely that it is associated with the specific loss of one or more chromosomes (554). However, reexpression may also occur at lower frequency in hybrids in which extinction is stable in the majority of the cells in the population, and in such cases it is commonly associated with chromosome loss (99,100,553). There are, however, no confirmed reports of the association of reexpression of a function with loss of an identified chromosome. Segregants from hepatoma-derived hybrids which reexpress liver-specific functions can be selectively isolated by the use of medium in which glucose is replaced as the carbon source by oxaloacetate or dihydroxyacetone, thus requiring activity of the gluconeogenetic pathway for growth (555,556). Reexpressing segregants may also be isolable on the basis of a shift in colony morphology or growth habit toward that of the expressing parent (219,557). Nevertheless, where a number of luxury functions are present in the expressing parent these may undergo independent reexpression (100). The phenomenon of reexpression allows two conclusions to be drawn about the mechanism of extinction. First, it is not due to structural gene loss, but must be due to the production by the nonexpressing genome of diffusible regulatory substances that directly or indirectly inhibit gene expression. Second, because functions may be reexpressed independently, extinction does not involve a change in the state of determination of the genome, but merely modifies the expression of individual characteristics of that state of determination. An observation that at first sight runs contrary to the expectation of independence of reexpression is seen in hybrids between Friend erythroleukemic cells and fibroblasts, where anchorage dependence correlates with inducibility for the expression of a wide spectrum of erythrocyte-specific differentiation products (557,558). However, the difference here is that these functions are not, like those of other cell lines commonly studied such as hepatomas, constitutively expressed by the Friend cell parent: What is expressed is the capacity to respond to an inducer by undergoing a specific developmental maturation (553).

In hybrids between melanoma and hepatoma cells extinction is reciprocal, and reexpression of functions characteristic of the two parental cell types is mutually exclusive ("phenotypic exclusion," ref. 559). In contrast, hybrids between Friend cells and lymphoma or myeloma cells produce surface antigens characteristic of the lymphoma or myeloma parent and are also inducible for hemoglobin synthesis (557). This difference in behavior may relate to the fact that in the latter case one is assaying a potential (see previous paragraph), or alternatively may relate to the degree of separation of the respective lineages: Melanoma and hepatoma cells are, respectively, of ectodermal (neural crest) and endodermal origin, whereas one is dealing in the other case entirely with hematopoietic cells.

4.3.3. Expression of Luxury Functions in Hybrids Between Cells of the Same Histiotype

The expression of luxury functions has been examined in hybrids between two different tumor-derived cell lines of the same histiotype but with differences in gene expression (159) and also between "de-differentiated" variants and their parental cell lines (508,556,560). Extinction is seen in some cases but not others, allowing conclusions to be drawn about the mechanistic basis of differences in expression as in dominance analysis of household functions. In hybrids between the rat hepatoma line Fao and the mouse hepatoma line BW1-J, which expresses a less extensive range of liver-specific functions, extinction was confined to those functions whose expression is normally activated neonatally, whereas functions expressed earlier in the fetus were not extinguished (159). In hybrids between a dedifferentiated hepatoma variant and its differentiated parent, transitory extinction was observed (556).

Hybrids may also be constructed between normal cells of finite lifespan and tumor cells of the same histiotype: Where extinction does not occur this serves to immortalize the pattern of gene expression of the normal cell type. The most important example of this approach is the production of hybrid myelomas or "hybridomas."

HYBRIDOMAS

In 1975 Kohler and Milstein (561) showed that spleen cells from an immunized mouse could be fused with myeloma cells, and that some of the resulting hybrids produced antibody directed against the antigen with which the mouse had been immunized. Such hybrids can be cloned and grown indefinitely, yielding potentially unlimited supplies of monoclonal antibody. This development has produced a revolution in immunology and the technology for producing hybridomas has developed rapidly. Myeloma cell lines suitable for producing hybridomas have now been developed from rat and human tumors as well as mouse (226,562). Although useful interspecies hybridomas have been produced (562) it is preferable that myeloma and immunized animal are of the same species unless there are specific reasons for this not to be so. This helps to avoid problems associated with chromosome loss, and also, in the case of mouse and rat hybridomas, allows high-titre antibody stocks to be produced by growing the hybridoma as an ascites tumor (226).

A number of factors govern the choice between rat and mouse: (a) One species may respond better, or more specifically, to the antigen; (b) rats are more convenient for growth of large amounts of ascites tumor; and (c) rat hybrids require more practice to derive but the percentage that produces the desired antibody is usually higher (226). Within each species one then has a choice of myeloma lines. Most myeloma lines themselves produce immunoglobulins (of unknown antigen specificity) and yield hybridomas in which the immunoglobulin chains of both myeloma and spleen cell parents are produced, and these associate to form mixed antibody molecules as well as the two parental types so that only a small fraction of the antibody has the antigen specificity of the spleen cell parent. This problem can be overcome by isolating segregants from the hybridoma by electrophoretic screening (226), but it is simpler to start with a nonproducing myeloma variant which yields hybridomas giving antibody of the spleen cell type only. Such variants of both mouse and rat myelomas are available, and nonsecreting mouse hybridoma segregants have also been used as fusion partners (226). These lines have been selected for HGPRT deficiency to enable hybrid selection in HAT medium, and residual spleen cells are readily eliminated from the fusion mixture as they fail to grow in culture. However, because of the heterogeneity of the spleen cell preparation only a small proportion of the hybrids so obtained produces antibody of the desired specificity, and a variety of rapid screening techniques are available to allow the desired hybrids to be isolated. Most depend on plating the fusion products in a large number of microwells and screening the culture supernatants, but techniques for screening single colonies using nitrocellulose filter replication or the hemolytic plaque assay (Section 2.3.1) have also been employed (226), as has fluorescence-activated cell sorting (217). Hybrid antibody molecules containing recognition sites for two different specified antigens can be obtained from "hybrid hybridomas" produced by fusing a hybridoma that synthesizes antibody to the first antigen with spleen cells of an animal immunized with the second antigen (200).

There is much interest in the development of human hybridomas for therapeutic purposes. Suitable human myeloma lines have been identified but nonproducing variants are not yet available (562). Alternatives to the use of spleen cells after *in vivo* immunization have been sought in order to circumvent ethical problems, and some success has been obtained with peripheral blood lymphocytes and with *in vitro* immunization techniques (562). The latter techniques may have advantages for mouse and rat hybridomas also, in that they are rapid, the conditions can be more closely controlled, and suppression and tolerance can be overcome to allow the production of antibodies to self-antigens and to antigens to which it is difficult to raise a response *in vivo* (562).

A similar approach is now being used to analyze T-lymphocyte function. Antigen-stimulated T cells are fused with cells of a T-cell lymphoma to yield T-cell hybridomas expressing T-cell functions such as the production of helper and suppressor factors and lymphokines (563–565). Here also antigen stimulation can be carried out *in vitro* with similar advantages to those applying to antibody production (566,567).

4.3.4. Expression of Tumorigenicity in Cell Hybrids

A vast and confusing literature exists regarding the behavior of tumorigenicity in cell hybrids (reviewed in refs. 100,568, and 569). There are a number of reasons for this confusion. First, the term "malignancy," which is correctly used to denote the capacity of a tumor to grow invasively, is often erroneously used to refer not only to tumorigenicity (the capacity to form a tumor of any kind) but also to any of a variety of phenotypes shown by cultured cells, such as insensitivity to density-dependent growth inhibition, which correlate imperfectly with tumorigenicity. Second, if a population of hybrid cells containing a low frequency of segregants that have undergone chromosome loss is injected into an animal, the formation of tumors by segregants may obscure lack of tumorigenicity in the cells that have not undergone chromosome loss, so that it is mandatory to examine the karyotype of tumors formed by injection of hybrids. Third, investigations have differed in the extent to which the immune response of the host animal to tumor antigens is taken into account: Even where immunosuppressed hosts have been used, the methods employed to achieve immunosuppression have differed. Fourth, parental cells have sometimes been of the same histiotype and sometimes not. Fifth, tumorigenicity studies have not always been sufficiently quantitative. Finally, in some cases tumorigenic cells of an established tissue-culture line carrying a selectable marker such as HGPRT deficiency have been fused to normal cells incapable of growth in culture, and hybrids have been selected on the basis of ability to grow in culture in selective media. This may have led to the preferential isolation of hybrids whose growth properties resembled that of the tumorigenic parent.

When one attempts to take such factors into account the following broad generalizations emerge (see refs. 100,568, and 569 for documentation).

First, hybrids between tumorigenic and nontumorigenic cells are usually nontumorigenic, although there may be exceptions. Nontumorigenic hybrids may yield tumorigenic segregants by chromosome loss. However, it has not been possible to correlate the regain of tumorigenicity with loss of an identified chromosome, and indeed extrachromosomal genes may also be involved: In one case hybrids which had lost all cytologically detectable chromosomes of the nontumorigenic parent remained nontumorigenic (570) and in another, cybrids formed by fusing cytoplasts from nontumorigenic cells with intact tumorigenic cells were not tumorigenic (571), although this does not appear to be a general phenomenon (572). Second, hybrids between two different tumorigenic cell lines are usually tumorigenic. This is at first sight surprising as one might expect a variety of complementing lesions to be capable of giving rise to tumorigeneicity. It is almost certainly due to the acquisition of multiple lesions by each line during the processes of tumor progression and establishment in culture which impose continual selection for rapid growth.

4.4. CHROMOSOMAL AND SUBCHROMOSOMAL GENE ASSIGNMENT

As mentioned in Section 4.3, chromosome loss is common in interspecific hybrids. This loss is unidirectional, that is, a given hybrid cell tends to retain a complete chromosome set from one parent and to segregate the chromosomes of the other. Furthermore, in a given fusion it is possible to predict the direction of segregation from a knowledge of the parental species and cell types. Early studies emphasized the role of the species: For instance, in hybrids between permanent lines of human and rodent origin it is the human chromosomes that are selectively lost. However, subsequent studies have shown that the nature of the cells used, particularly whether they are diploid cells or cells of an established line, is at least as important as the species and indeed can override it: For instance, hybrids formed from mouse primary culture cells and established human cells segregate mouse chromosomes (100). Pravtcheva and Ruddle (573) have suggested a role for a cellular system recognizing native and foreign X chromosomes in determining the direction of chromosome loss. They found that while hybrids between established Chinese hamster and mouse lines segregated Chinese hamster chromosomes, if the Chinese hamster parent contained a single mouse X chromosome introduced by microcell fusion then its hybrids with established mouse lines segregated mouse chromosomes provided that the mouse line used to construct the final hybrid differed from that from which the X chromosome originated. If the same mouse line was used in both steps, the hybrid segregated Chinese hamster chromosomes and retained those of mouse origin, including both X chromosomes. The X chromosome of a diploid cell was as effective as that of a permanent cell line in reversing the direction of chromosome loss (574). If it is desired to produce hybrids segregating chromosomes of a particular mammalian species, then the best strategy is to fuse diploid lymphoblasts or fibroblasts of that species to cells of an established rodent line. This has been successful for a wide variety of both rodent and nonrodent species (see 575 for references). This may be an example of a more general rule that it is usually the chromosomes of the slower-growing parent which are selectively eliminated (100), although exceptions do exist (576). It is an interesting question why chromosome loss should be unidirectional at all. One factor that appears to be important is incompatibility between the mitochondria of one parent and the nucleus of the other. There is rapid elimination of all the mitochondria of the parent whose chromosomes are preferentially lost (263). It is not possible to isolate hybrids retaining the mitochondria of the segregating parent by applying selective techniques after chromosome segregation has begun. However, by selecting immediately after fusion for retention of human mitochondria it is possible to reverse the direction of both mitochondrial and chromosomal segregation from mouse–human hybrids (263).

Segregation of chromosomes of the diploid parent from hybrids between diploid cells and cells of established rodent lines has provided a powerful method for the assignment of genes to chromosomes. Any gene whose product can be distinguished from the homologous gene product of the fusion partner can be assigned to a chromosome on the basis of coordinate segregation of the chromosome and the gene product (577). Once a gene has been assigned to a chromosome its product can be used as a marker to speed further assignments. Particularly useful for this purpose have been isozyme markers (577,578). Proteins with defined mobilities on two-dimensional polyacrylamide gel electrophoresis may also prove valuable (579), and the direct detection of DNA sequences by Southern blot techniques makes it possible to extend the method to genes whose products are not expressed (577).

Two markers that cosegregate are described as syntenic; this term is not synonymous with "linked" which denotes a meiotic recombination frequency of less than 50%. Since the total length of the human genome is some 3000 centimorgans (cM), it is possible for widely separated syntenic markers to show no linkage. Markers can be ordered within a synteny group and assigned to regions of chromosomes by using parental cells carrying defined translocations. The limit of resolution of this technique is imposed by the precision with which translocation endpoints can be defined using chromosome banding techniques (Section 1.6) and is of the order of $5-10$ cM, that is, $(5-10) \times 10^6$ base pairs in human cells (580).

An alternative technique for the assignment of genes to chromosomes is *in situ* hybridization of nucleic acid probes to metaphase spreads (577,581). Until recently, for reasons of sensitivity, this could only be applied to reiterated sequences, but a number of recent developments have brought single-copy genes within the scope of the technique. These include the use of high specific activity ^{3}H- or ^{125}I-labeled probes, the use of single-strand probes to avoid competing self-annealing, an increase in the length of the probe molecule beyond 5 kb, and the use of dextran sulfate which increases sensitivity, possibly by promoting the formation of probe clusters at the site of hybridization (577,581). Typically with such techniques the extent of labeling of single-copy genes is of the order of one silver grain per chromosome, so that one needs to examine a large number of spreads to achieve a statistically valid assignment. With ^{125}I-labeled probes the resolution is limited by the path length of the radiation, but high-resolution localization comparable to that obtainable by translocation mapping can be obtained by the use of ^{3}H-labeled probes in conjunction with banding techniques which can be applied after autoradiographic exposure, such as the BUdR/Hoechst 33258/Giemsa staining method described in Section 1.6 (98 and references cited therein).

A considerable amount of mapping data has now been obtained for the human genome, and mapping of the genomes of other mammals has begun but is at a less advanced stage (242,575,577,582). Comparison between the synteny groups of different mammalian species reveals that the X chromosome is conserved as a synteny group throughout: This was predicted by Ohno (583) because disruption would upset dosage regulation. Perhaps less predictable is the degree of conservation found for autosomal synteny groups: Those of human and chimpanzee are identical except that human chromosome 2 corresponds to two acrocentric chromosomes in the chimpanzee; the majority of synteny groups are conserved between human and cat, and only about half are disrupted between human and mouse, mostly due to breaks at or near centromeres (577,582). Also of interest is the arrangement of the elements of multigene families in the human genome: While the elements of some families are clustered on a single chromosome, others are widely dispersed (584).

4.5. CYBRIDS

Cells in monolayer culture can be enucleated by treatment with cytochalasin B and centrifugation (585), when the nucleus surrounded by a small amount of cytoplasm ("karyoplast" or "minicell") is extruded and migrates in the centrifugal field leaving the remaining cytoplasmic fragment ("cytoplast") attached to the substratum. The resulting cytoplast preparation is usually contaminated with whole cells but may be partially purified by differential trypsinization or the use of Ficoll gradients. Enucleation of suspension cells can be achieved by treatment with cytochalasin B and centrifugation of the whole treated preparation in a Ficoll gradient. Alternatively, modified substrates may be used to allow monolayer enucleation of poorly adherent cell types (585). When cytoplasts are fused with whole cells the initial products, termed "fusion cybrids," are not all capable of further division, but in general proliferating "viable cybrids" can be isolated from them at a higher frequency than that with which viable hybrids are obtained from heterokaryons, presumably because no problems of nuclear asynchrony arise (572). In devising selective systems for viable cybrids one must bear in mind that the cytoplast preparation is contaminated with residual intact cells, so that one selects for the nucleus of the intact partner and for the cytoplasm of the enucleated partner but against its nucleus. For instance, cytoplasts from an $HGPRT^+CAP^R$ cell may be fused to intact $HGPRT^-CAP^S$ cells and cybrids selected in thioguanine plus chloramphenicol (572). Alternatively, cytoplasts can be rendered fluorescent either by allowing them to ingest fluorescent beads or by incubating them in a fluorescent lipid derivative which is incorporated into the membrane, and cybrids selected using the fluorescence-activated cell sorter (586).

Study of cybrids has been important in establishing cytoplasmic inheritance of mitochondrial DNA-coded variant phenotypes such as chloramphenicol resistance (Section 3.5.6) and also of intracisternal A particles and microtubule-organizing centers (572). Cytoplast fusions have also been used to investigate the role of cytoplasmic factors in the regulation of differentiation and tumorigenicity. However, no clear rules have emerged from such experiments: Depending upon the cell types used one may observe stable or transitory extinction of luxury functions of the intact cell parent, stable or transitory activation of luxury functions characteristic of the cytoplast parent, or no effect (553,572). Similarly, although instances of stable suppression of tumorigenicity by cytoplast fusion are known this is not a general phenomenon (Section 4.3.4).

4.6. KARYOPLAST AND MICROCELL FUSIONS

As prepared by the technique of centrifugation in cytochalasin B (Section 4.5) karyoplasts contain, on average, about 10% of the cytoplasm of the cell, and are contaminated with whole cells and small cytoplasmic fragments. The karyoplasts may be purified by gravity sedimentation or differential adhesion, but the first method does not yield very pure preparations and the second requires time and is accompanied by deterioration (587). It is more satisfactory either to separate them using a fluorescence-activated cell sorter, on the basis of light scatter (587) or after staining with a fluorescent mitochondrial stain such as rhodamine 123 (588), or to allow the cells to ingest particles of tantalum prior to enucleation in order to increase the difference in sedimentation rate between intact cells and karyoplasts. The latter technique permits subfractionation of the karyoplasts on the basis of cytoplasmic content, so that preparations containing only 2–4% of the cell cytoplasm can be obtained (587). Such karyoplasts are incapable of regenerating the lost cytoplasm to form viable cells but after fusion with cytoplasts yield viable products that are termed "reconstructed" or "reconstituted" cells (587). Reconstituted cells may be selectively isolated by techniques similar to those used for cybrids (Section 4.5), and allow one to study the effect of the recipient cytoplasm on the donor nucleus without much dilution by donor cytoplasm. However, in interspecific fusions incompatibility is observed between the karyoplast nucleus and the cytoplast mitochondria as is the case for whole cell hybrids, and the mitochondria of the karyoplast parent repopulate the cytoplasm of the reconstituted cell with the selective elimination of those of the cytoplast parent (263). Karyoplasts may also be fused with whole cells to give "nuclear hybrids" or "karyobrids" (588).

In general, the properties of karyobrids are similar to those of whole cell hybrids, except that in some interspecific fusions a higher yield of viable colonies is obtained from the initial fusion products in the case of karyobrids than is the case with whole cell hybrids, presumably because the relative numbers of mitochondria of the two parents are more compatible with long-term survival (588).

When rodent cells are treated with mitotic arrest agents such as colcemid for 36–48 hr they undergo micronucleation, and upon centrifugation in cytochalasin B the micronuclei are extruded as separate membrane-bounded bodies termed "microkaryoplasts" or "microcells," the smallest of which contain only a single chromosome (589). An alternative technique to achieve micronucleation without prolonged mitotic arrest is to collect preparations enriched in mitotic cells by selective detachment after brief colcemid treatment and to replate them in the presence of cytochalasin B (589). These techniques do not work well with human cells, but microcell-like fragments can be generated by colcemid treatment followed by incubation in colcemid-free medium (590) or by exposure of thymidine-arrested cells to hyperbaric nitrous oxide followed by cooling to 4°C ("minisegregants," ref. 591). Alternatively, microcells containing human chromosomes can be obtained from human–rodent hybrids by techniques used for rodent cells (592). Purification of microcells can be achieved by membrane filtration or gravity sedimentation (589). Fusion of microcells to intact cells is a useful technique for the isolation of hybrid cells containing a small number of intact chromosomes of one parent. Opinions differ regarding the relative merits of polyethylene glycol and Sendai virus in mediating microcell and karyoplast fusions (589,590,593). Perhaps the most useful microcell hybrids are those containing a single donor chromosome maintained by selective pressure: The relative stability and homogeneity of the donor chromosome content, as well as its simplicity, makes such hybrids invaluable for chromosomal gene assignment.

4.7. HIGHER-RESOLUTION MAPPING TECHNIQUES

As stated in Section 4.4 the limit of resolution of translocation mapping is in the region of 5–10 cM, while it is currently difficult to cover more than about 0.1 cM of the genome (10^5 base pairs) by recombinant DNA techniques alone (580). Two techniques have been employed to bridge the gap between

0.1 and 10 cM. Goss and Harris (594,595) carried out cell fusions in which one partner had been X-irradiated to introduce chromosome breaks, and selected for the retention of a chosen marker in the resulting hybrids. The proportion of such hybrids which retained a second, unselected marker carried on the same chromosome was then a function of its distance from the selected marker. By calibration using genes whose separation was known from translocation mapping they were able to determine the distance between the human thymidine kinase and galactokinase genes, which map to the same band of chromosome 17, as 1.2 cM. The frequency of disruption of a linkage is simply related to the X-ray dose, so that the resolution of the technique can be varied. The other technique which has been used to bridge the resolution gap is that of chromosome-mediated gene transfer (596,597). In this technique recipient cells are incubated in a suspension of metaphase chromosomes isolated from the donor cell type. Chromosomes are taken up by phagocytosis, a process which is increased in efficiency by use of calcium phosphate coprecipitation and postincubation of the cells in DMSO (compare Section 5.1). However, most of the ingested chromosomal DNA is degraded to small fragments, so that the frequency of transfer of any chosen marker is typically in the range 10^{-5}–10^{-7} and selective methods are thus necessary to isolate the resulting transformants. Free functional chromosomal fragments, which may be as large as 1% of the haploid genome and which may be present either as single or as multiple copies, can be maintained for long periods in the transformants if selection is continued, but if selection is relaxed they are lost at the rate of about 1–10% per generation. However, the foreign chromosome segment or "transgenome" can become stably established by a low-frequency secondary event. Klobutcher and Ruddle (598) using alkaline Giemsa staining, which distinguishes human from mouse chromosomes (Section 1.6) were able to show that stabilization of a human transgenome in mouse cells involved integration of the transgenome into nonhomologous sites on mouse chromosomes, and was accompanied by a decrease in size of the transgenomic fragments. Chromosome-mediated gene transfer thus provides two parameters related to the distance separating a chosen pair of markers: first, the proportion of those primary (unstable) transformants selected for one marker which also carry the other, and second, the proportion of independent stabilization events in which an association present in an initial unstable transformant is disrupted. Klobutcher and Ruddle (598) were able to show by banding that different stabilization events occurring in a single unstable transformant all involved deletion of material from the centromere-proximal end of the fragment, enabling them to order markers relative to the centromere. A problem which must be borne in mind when using chromosome-mediated gene transfer, however, is that multiple donor fragments may be present in the recipient cells and in some cases these can become associated (598) making it mandatory to examine a number of independent primary transformants.

5

MOLECULAR SOMATIC CELL GENETICS

A number of techniques are now available which allow the incorporation of purified DNA into cultured mammalian cells (Section 5.1). DNA coding for a chosen marker can be successfully incorporated using total genomic DNA as source if the marker can be efficiently selected (Section 5.2.1). This approach has been spectacularly profitable in demonstrating the existence of oncogenes in the DNA of certain tumors (Section 5.5). More efficient incorporation can be achieved if the chosen sequences are first purified from cellular DNA by molecular cloning (Section 5.2.2), and a range of vectors has been developed specifically for this purpose (Sections 5.2.3–5.2.6). Some of these allow the inserted DNA to be maintained in the recipient cells as an episome while others allow its integration into host chromosomes with the minimum of rearrangement (Section 5.2.3). Certain vectors are so constructed as to allow the expression of any coding sequence inserted into them and can be used to ensure the expression either of cDNA copies obtained by reverse transcription of mRNA, or of bacterial genes (Section 5.2.4). Either bacterial genes or cDNA, cloned in such an expression vector, or genomic DNA may provide a selectable marker (Section 5.2.5). The DNA segment of interest may be subjected to directed *in vitro* mutagenesis prior to its introduction into recipient cells, and this is beginning to provide a further route for the introduction of genetic markers (Section 5.4).

5.1. TECHNIQUES FOR THE INTRODUCTION OF DNA INTO MAMMALIAN CELLS

Purified DNA may be incorporated into tissue-culture cells by transfection, by microinjection, or by the use of liposomes. Transfection involves incubating the cells with DNA either in the presence of DEAE-dextran (599) or in the form of a calcium phosphate coprecipitate (600). The latter technique is more efficient, particularly if the cells are posttreated with DMSO (601) or glycerol (602). As with chromosome-mediated gene transfer the majority of the DNA taken up suffers degradation, so that the frequency of transfection for a specified marker is again usually of the order of 10^{-6}, necessitating selective techniques for the isolation of the transfectant. The presence of carrier DNA (which should not itself carry the selectable marker) is important to obtain good transfection efficiency: There is an optimal concentration of carrier DNA in the region of 10 μg/mL. It appears to influence both the morphology of the coprecipitate and the fate of the DNA once taken up. In the presence of carrier DNA a higher proportion of primary transfectants is unstable in the absence of selective pressure, and the marker DNA of the unstable transfectants is in high-molecular-weight DNA which, nevertheless, is smaller than intact chromosomes. It seems reasonable to conclude that ligation of the marker DNA has occurred to the carrier DNA, and that this process aids maintenance of the marker DNA either by protecting it against exonuclease attack or by virtue of the presence of elements such as replication origins in the carrier DNA (603). Secondary stabilization events subsequently occur at low frequency, involving integration into chromosomes (compare Section 4.7). However, even "stable" transgenomes appear to be more susceptible both to loss and to amplification than native genes (603). As a result of the intracellular ligation process it is possible to isolate transfectants for nonselectable markers simply by cotransfection with DNA carrying a selectable marker: Prior covalent linkage between the two markers is unnecessary (603). This widens considerably the range of application of the technique. Not all cell types, however, are equally susceptible to transfection. The process occurs quite efficiently with L cells but with many other cell types transfection frequencies are appreciably lower (604,605).

Microinjection into tissue-culture cells of volumes of DNA solutions in the range 10^{-10}–10^{-11} mL can be achieved using a 0.5-μm diameter glass capillary operated by gentle air pressure from a syringe. It is possible to inject into the cytoplasm, the nucleus, or even a single mitochondrion (606). By microinjection into the nucleus it is possible to achieve frequencies of stable incorporation of a desired marker as high as 20% (603), making a selective system unnecessary. A simpler technique in which cells are overlaid with saline containing the appropriate DNA and then pricked over the nucleus with a microneedle gives a frequency of stable incorporation of about 2% (607). As with transfection, coinjection of a mixture of DNA molecules leads to a high frequency of coincorporation (603). Microinjection by iontophoresis rather than pressure leads to multiple integration without the tandem insertions commonly seen in pressure-injected cells. Because there is no net fluid displacement, shearing is minimized and very large amounts of DNA can be injected. More nucleolytic degradation of the injected DNA occurs prior to integration but it is nonetheless possible to obtain integration of an injected restriction fragment in its entirety (608).

DNA molecules may be encapsulated within the aqueous interior of liposomes, or phospholipid vesicles, and thence delivered into cells (609). The efficiency of delivery may be increased by the use of dimethyl sulfoxide, ethylene glycol, glycerol, or polyethylene glycol, and also by incorporation into the liposome membrane of glycolipids, lectins, or covalently bound antibodies to increase binding to the cells or even to confer specificity for a particular cell type in a mixture should this be desirable (609). Polyethylene glycol-stimulated uptake is insensitive to a wide range of metabolic inhibitors and probably proceeds by direct membrane fusion, while stimulation by glycerol is quite sensitive to metabolic inhibition and is enhanced by agents such as chloroquine which are selectively taken up into lysosomes (lyso-somotropic agents), suggesting that glycerol acts by promoting endocytosis (609).

DNA in the form of a plasmid (or cosmid) can be incorporated into tissue-culture cells by fusion of bacterial protoplasts to the cells using polyethylene glycol (610,611). This avoids the necessity for purification of the plasmid DNA after growth in its bacterial host and its efficiency compares favorably with that of calcium phosphate-mediated transfection. As with transfection, integration of the transgenome appears to occur by the formation of con-catemers, and cotransfer of markers carried on separate plasmids can be achieved by fusing a mixed population of bacterial cells carrying the respective plasmids to the mammalian cells (unpublished results cited in ref. 612).

The degree of success achieved in the transfer of markers using mito-chondrial DNA has been variable: Both positive (613) and negative (614) results have been reported. However, in some of the latter cases markers could be successfully transferred by incubation of cells with isolated mi-tochondria (614).

Other techniques used for the introduction of macromolecules into mam-malian cells have not yet been successfully applied to DNA. Using erythrocyte ghost fusion it appears difficult to introduce sufficiently large molecules into erythrocytes during hypotonic hemolysis (615). A relatively new tech-nique, that of osmotic lysis of pinocytic vesicles, has as yet been applied only to proteins and carbohydrates (616).

5.2. SOURCES OF DNA FOR INTRODUCTION INTO MAMMALIAN CELLS

5.2.1. Genomic DNA

It is possible to obtain successful transfection for a chosen marker simply by the use of total genomic DNA from cells carrying the marker without any prior purification of the appropriate DNA sequence, provided an efficient selective system is available. Transfection frequencies using genomic DNA for TK and APRT are usually appreciably higher than with HGPRT (603): This is due to the greater length of the *hgprt* gene which makes breakage of the DNA more of a problem (617). It is possible to achieve transfection for a purified gene coding for a nonselectable marker by cotransfection with genomic DNA (618), but cotransfection with a selectable marker cloned in an expression vector (Sections 5.2.4–5.2.6) is generally more efficient. Transfection with genomic DNA has been particularly important in the study of oncogenes (Section 5.5).

5.2.2. Cloned DNA Sequences

The sequences that one wishes to introduce into mammalian cells may be purified from total genomic DNA by molecular cloning. A comprehensive review of techniques for the construction and screening of libraries of genomic DNA clones is beyond the scope of this book and the reader is referred to ref. 619. Particularly worthy of mention, however, is a technique that has been used for the cloning of chicken *tk* and hamster *aprt* genes (620,621) and is potentially applicable to any marker selectable in mammalian cells. In the first example, high-molecular-weight fragments of chicken DNA were ligated to DNA of the plasmid pBR322 so that all chicken sequences, including the *tk* gene, became covalently linked to a pBR322 sequence. The DNA preparation thus obtained was used to transfect LTK⁻ cells. By selecting for TK⁺ colonies, transfectants were obtained with integrated *tk* genes flanked by pBR322 sequences. DNA from these primary transfectants was then used to transfect more LTK⁻ cells to TK⁺, thereby reducing the level of pBR322 sequences unlinked to *tk*. The DNA of the secondary transfectants was then digested with a restriction enzyme which does not cut the *tk* gene but cuts once within pBR322. It was then ligated to allow cyclization and used to transform *E. coli*, selecting for the antibiotic resistance marker carried by pBR322. The resulting plasmid clones were then screened for ability to transfer *tk* to LTK⁻ cells by transfection.

In order for this strategy to be successful the fragment of chromosomal DNA contained in the final plasmid clone must contain an entire transcription unit. This makes its application difficult in cases where the transcription unit is very long, for example, dihydrofolate reductase (>42 kb in the mouse, ref. 622). An alternative strategy is to provide only the sequences of the mature mRNA in the form of a cDNA copy and to incorporate processing signals into the vector to allow its expression: A similar strategy can be used to ensure the expression of bacterial genes in mammalian cells (Section 5.2.4). A further limitation of the plasmid clones described above is that they are not capable of replication in mammalian cells so that, as with genomic DNA, in order for a marker to be stably retained it must undergo integration into a chromosome. The integration process and the ligation processes which precede it (Section 5.1) can interfere with gene expression as is seen in the case of DNA sequences introduced by techniques that do not select for their expression, for example, cotransfection and microinjection. There is considerable heterogeneity in expression between independently isolated clones, sometimes associated with improper initiation of transcription, which is at least in part due to sequence rearrangements occurring prior to or at the time of integration (605). In some cases such rearrangements can continue long after the initial integration event, leading to heterogeneity of expression between different cells of the same clone (623). Therefore, considerable effort has been devoted to the development of vectors capable of multiplication in mammalian cells (Section 5.2.3).

5.2.3. Vectors Capable of Multiplication in Mammalian Cells

One approach to the construction of vectors capable of multiplication in mammalian cells is to construct viral derivatives which can propagate as infectious virions (623,624). This, however, imposes a size limitation because of the necessity for packaging into the virion particle. Furthermore, no well-characterized mammalian virus has a large nonessential region that can be replaced by foreign DNA, and, therefore, in vectors of this kind which have been constructed the inserted DNA replaces essential genes and the recombinant genome is defective and requires a helper virus for propagation, generating a mixed stock. However, in some cases the helper function can be incorporated into the host cell genome. Because it is tedious and expensive to carry out the primary cloning of the DNA sequence in mammalian cells this is usually done in *E. coli* by first cloning the viral vector in a plasmid and then inserting the foreign DNA. The recombinant viral genome is then excised from the *E. coli* plasmid vector and introduced into mammalian cells. As no *in vitro* packaging systems are available for animal viruses, this must initially be achieved by a technique such as transfection.

Among such virion vectors those based on retroviruses appear to offer particular promise (624). These single-stranded RNA viruses are converted within the cell to a double-stranded proviral DNA via a reverse transcription step in which their termini undergo duplication to produce long terminal repeats (LTRs). The LTRs contain promoter and polyadenylation sequences (Section 5.2.4) and determine the specificity of proviral integration: The provirus can integrate at many chromosomal sites but always does so by recombination at defined sites at the end of the LTRs. Thus, in contrast to the behavior of papovavirus vectors, one can be confident that sequences inserted between the LTRs will be integrated in an intact state. Other advantages offered by retroviral vectors are that the multiplication of the virus does not kill the cell, and that they have a wide host range which can be extended by pseudotyping, that is, packaging the genome in the coat of a related but different retrovirus, which is achieved by mixed infection. As with the naturally occurring strongly transforming retroviruses, the inserted sequences in retrovirus vectors replace genes required for replication. The vectors can be either propagated with a helper virus or cloned in an *E. coli* plasmid or bacteriophage λ vector and then introduced into recipient cells by transfection. The recombinant virus can then be rescued from the integrated state by superinfection with helper virus. The cis-acting functions required for retrovirus multiplication are all located in or close to the LTRs, making it possible to construct helper cell lines which provide the trans-acting *gag*, *pol*, and *env* gene products and allow the growth of defective retrovirus vectors in which these genes are replaced by the inserted DNA (625, 626). Although retroviruses employ RNA splicing to produce subgenomic mRNAs they do not contain introns within the coding sequence of individual genes (627), and when genes containing introns are cloned in retrovirus vectors the introns are excised from the virion RNA (624).

Virion vectors based on other viruses, however, suffer from the disadvantage that the multiplication of the virus kills the host cell. Therefore, efforts have been made to develop systems in which the recombinant DNA replicates as an episome, conferring the dual advantages of producing stable cell lines and of freedom from packaging constraints.

Two approaches have been used to develop vectors which replicate episomally. Gluzman (628) isolated a line of cells designated COS by transforming the monkey cell line CV-1 with a replication origin-defective early region fragment of SV40 DNA. The resulting cells express SV40 T antigen and contain the permissivity factors for SV40 DNA replication. Indeed, the only cis-acting function required for SV40 DNA replication in COS cells is a segment of approximately 85 base pairs surrounding the replication origin. Plasmids containing this small segment of the SV40 genome will replicate in COS cells, although the efficiency of replication depends upon the plasmid vector used. In particular, the plasmid pBR322 contains sequences (so-called "poison sequences") which inhibit replication and whose deletion results in a marked increase in replication efficiency. Efficient amplification and, in some cases, expression of genes cloned into suitable plasmid vectors containing the SV40 origin have been achieved after transfection of COS cells. However, no permanent lines expressing the exogenous gene have yet been obtained, perhaps because the cells cannot tolerate the presence of such high levels ($2-4 \times 10^5$ copies per cell) of replicating episomal DNA (623). A further problem with such vectors is that passage in mammalian cells leads to accumulation of a large number of mutations, including point mutations, deletions, and duplication/insertions (629,630).

The second approach makes use of the fact that transformation of rodent cells by bovine papillomavirus (BPV) is not accompanied by detectable levels of integration of viral sequences, the viral genome being maintained as an episome. A fragment comprising 69% of the viral genome retains the ability for *in vitro* transformation and has been used as a vector for introduction of the rat preproinsulin gene into mouse cells, using morphological trans-formation as the selective marker. Stably transformed cells result which maintain the recombinant genome as an episome present at about 100 copies per cell and efficiently express the rat preproinsulin gene (623). This vector is propagated as a plasmid clone in *E. coli*, but the plasmid sequences must be removed prior to transformation because they inhibit replication in mam-malian cells (see preceding paragraph). This problem has been circumvented recently by the construction of shuttle vectors in which the BPV transforming fragment is linked to a pBR322 fragment lacking poison sequences (Section 5.2.6). As is the case with the SV40-based vectors discussed above, some BPV-based vectors are unstable: This appears to depend upon the detailed construction of the vector, and, in particular, on the nature of the selectable marker used (Section 5.2.6).

5.2.4. Expression Vectors

In addition to the coding sequence a number of processing signals are nec-essary to ensure expression of a gene in a mammalian cell (623,631). Initiation of transcription requires a promoter region which, in the case of genes transcribed by RNA polymerase II, consists of the so-called "TATA box" (a sequence $TATA(^A/_T)A(^T/_A)$ or one closely related to this) about 30 nucleotides upstream of the site of transcription initiation, plus other sequences further upstream, sometimes including elements termed enhancers which boost the level of expression. The presence of introns and their removal by splicing appear to be obligatory for the production of stable cytoplasmic transcripts of certain genes. Thus although some eukaryotic genes do not contain introns at all and others appear to function normally if their introns are removed, the strategy employed for the construction of expression vectors has been to assume the necessity for splicing unless the converse is demonstrated for the gene in question (623). Finally, the sequence AAUAAA, found close to the 3' end of polyadenylated mRNAs, is required for the cleavage step which precedes polyadenylation (632).

Expression vectors have been constructed in which promoter, splicing, and polyadenylation sequences are provided in a configuration that ensures the expression of an inserted DNA fragment. Because viral transcription units are among the best characterized they have formed the basis of many expression vectors: Particularly well exploited have been the herpes simplex virus (HSV) *tk* transcription unit (605), the early and late transcription units of SV40, and the processing signals present in retrovirus LTRs (Section 5.2.3). Avian sarcoma virus (ASV) LTRs have the advantage of providing promoter function both in mammalian cells and in bacteria (633,634). Promoters of cellular genes such as that for metallothionein-1 (*mt-1*) have also been used in expression vectors (635). The *mt-1* promoter and the promoter present in the LTR of mouse mammary tumor virus (MMTV) have the advantage that transcription from them can be regulated, in the former case in response to the concentration of heavy metal ions such as Cd^{2+} (636) and in the latter by glucocorticoids (623).

5.2.5. Vectors Carrying Selectable Markers

The *tk* gene of HSV has been introduced into a number of plasmids as a complete transcription unit and serves as a selectable marker but the requirement for TK^- recipient cells restricts its application (605). By the use of a modified HAT selection with a high concentration of folate antagonist (methotrexate) and a low concentration of thymidine, Mercola et al. (637) were able to select for HSV*tk* expression in wild-type cells, but it remains to be seen whether these selective conditions are stringent enough to have general applicability. Mouse dihydrofolate reductase (*dhfr*) cDNA (623) and the *E. coli* galactokinase gene (638) have been cloned in expression vectors but likewise require mutant (DHFR⁻ or galactokinase-deficient) recipient cells to be useful as selectable markers. The use of *dhfr* cDNA does have the advantage, however, that once the recombinant genome has integrated into a recipient chromosome it is possible by selecting for methotrexate resistance to obtain amplification of the *dhfr* gene together with adjacent sequences (see Section 3.5.4), so that by inserting a nonselectable gene adjacent to the *dhfr* gene in the vector one can achieve amplification of the nonselectable gene following integration (623).

Dominant selectable markers that do not require the use of variant cells as recipients have also been developed. The *E. coli gpt* gene which encodes the enzyme xanthine guanine phosphoribosyl transferase (XGPRT), the bacterial analogue of HGPRT, has been cloned in expression vectors and serves as a selectable marker in wild-type recipient cells by virtue of its ability to use xanthine as a substrate. By using mycophenolic acid to block the conversion of IMP to XMP and supplying exogenous xanthine, cells can be made dependent upon *gpt* expression for survival. The stringency of this selection can be increased by additionally blocking IMP synthesis with aminopterin, in which case one must also supply adenine (or hypoxanthine) and thymidine (639). Alternative selective systems are possible if one uses HGPRT⁻ recipient cells: With some HGPRT⁻ cell types normal HAT medium will allow the growth of cells expressing *gpt* (640), although in other cases this apparently leads to adenine starvation and adenine must also be supplied (640). Mycophenolic acid plus guanine also constitutes an efficient selection for *gpt* in HGPRT⁻ cells (641).

A second dominant selective system employs the bacterial gene for aminoglycoside 3' phosphotransferase (*neo*) which originates from the transposon Tn5 and confers on bacteria resistance to neomycin and kanamycin. When cloned in an expression vector and introduced into mammalian cells this gene confers resistance to G-418 (geneticin), a 2-deoxystreptamine antibiotic which inhibits eukaryotic protein synthesis (623). A third dominant selective system employs the gene for a bacterial dihydrofolate reductase intrinsically resistant to methotrexate (623). Finally, it is possible to select for cells carrying multiple episomal copies of the human metallothionein-II$_A$ (*mt-II$_A$*) gene carried on a bovine papilloma virus-based shuttle vector on the basis of their increased resistance to cadmium salts (642).

5.2.6. Survey of Vectors Useful for the Introduction of DNA into Mammalian Cells

Tables 5.1–5.4 list some of the more useful vectors for DNA-mediated gene transfer in mammalian cells. It should be emphasized that the construction of vectors is a rapidly evolving field, and it is certain that new vectors with improved properties will have been developed before this book is published. Berg and his collaborators have developed a series of vectors (Table 5.1) which contain an *E. coli* plasmid vector derived from pBR322 (see ref. 619) and processing signals, generally from SV40 transcription units, placed to ensure the expression of an inserted foreign gene. In some cases an intact SV40 early region is also present, providing both a replication origin and trans-acting factors, notably T antigen, to allow replication in cells permissive for SV40, that is, monkey cells. In pSV5-X an intact early region from the related polyoma virus is present, allowing replication in mouse cells. In other vectors only the replication origin of SV40 is present, restricting replication to COS cells (Section 5.2.3) unless a helper virus is present. Even where episomal replication is possible, however, its efficiency is reduced by the poison sequences of pBR322 (Section 5.2.3) which are present in all the vectors of Table 5.1. Furthermore, the generation of permanent cell lines stably expressing the foreign gene requires integration into a chromosome of the recipient cell which is frequently accompanied by sequence rearrangement (Section 5.2.3). The inserted gene X may be one of the selectable markers discussed in Section 5.2.5. Another useful insert is the chloramphenicol acetyltransferase gene (*cat*): While an efficient selective system has yet to be devised for the expression of this gene, it has the advantage of a very sensitive assay with negligible background activity in mammalian cells (644).

The vectors pSVM-X and pMTV-X allow transcription of the inserted gene from the regulated promoter in the LTR of mouse mammary tumor virus, and in pSV*tk*-X and pSV*tk*-Xβ transcription occurs from the herpes simplex virus *tk* promoter, which is advantageous in embryonal carcinoma cells where the efficiency of the SV40 early promoter is low (Section 6.1.8). In three of these vectors a DNA segment encompassing the SV40 early promoter is also present, and contains enhancer sequences that augment the level of transcription from the MMTV or HSV*tk* promoter. The mechanism of the enhancer effect is not known, but it is cis-acting, independent of the orientation of the enhancer sequence relative to the gene, and can be prop-

TABLE 5.1. SV40-BASED EUKARYOTIC EXPRESSION VECTORS

Vector[a]	Eukaryotic Replication		Processing Signals for Expression of Inserted DNA(X)			Inserts (X) Available[c]	References
	Origin	Trans-Acting Factor(s)	Promoter	Splice Site	PolyA[b]		
pSV1GT5-X	SV40	SV40	1[d]	16s[e], 19s[f]	1	gpt* neo* dhfr*	643, 660, 661
pSV1GT7-X	SV40	SV40	1	19s	1	gpt* neo* dhfr*	643, 660, 661
pSV0-X	—	—	—	t[g]	e[h]	gpt* neo* dhfr*	643, 660, 661
pSV2-X	SV40	—	e	t	e	gpt* neo* dhfr* cat βG galK*	638, 643, 644, 660–662
pSV3-X	SV40	SV40	e	t	e	gpt* neo* dhfr*	660, 661
pSV5-X	SV40 + Py[i]	Py	e	t	e	gpt* neo* dhfr*	660, 661
pSVβ-X	SV40	—	e	βGL[j]	e	neo*	641
pSVM-X	SV40	—	e + MMTV[k]	t	e	gpt* dhfr*	647, 663

pMTV-X	—	MMTV	t	e	*dhfr**	647
pSVtk-X	SV40	e + tk[l]	t	e	*gpt**	641
pSVtk-Xβ	SV40	e + tk	βGL	e	*neo**	641
pCD-X	SV40	e	16s, 19s	l	See text	648

[a] X denotes inserted foreign DNA.
[b] Polyadenylation site.
[c] Selective markers denoted by *. *gpt*, *E. coli* xanthine guanine phosphoribosyltransferase gene; *neo*, *E. coli* aminoglycoside 3′ phosphotransferase gene; *dhfr*, dihydrofolate reductase cDNA; *cat*, bacterial chloramphenicol acetyltransferase gene; βG, rabbit β-globin cDNA; *galK*, *E. coli* galactokinase gene.
[d] From SV40 late transcription unit.
[e] Splice sites involved in production of SV40 16s mRNA.
[f] Splice sites involved in production of SV40 19s mRNA.
[g] SV40 small t-antigen mRNA splice sites.
[h] From SV40 early transcription unit.
[i] Polyoma.
[j] Rabbit β-globin large intron splice sites.
[k] Mouse mammary tumor virus LTR (promoter regulated by glucocorticoids).
[l] Herpes simplex virus thymidine kinase promoter.

TABLE 5.2. OTHER PLASMID VECTORS USEFUL FOR INTRODUCTION OF DNA INTO MAMMALIAN CELLS

Vector	Eukaryotic Replication System	pBR322 "Poison" Sequences	Selectable Marker	Processing Signals for Selectable Marker			References
				Promoter	Splice Sites	Polyadenylation Site	
pBPVβ1, pBPVβ3	BPV[a]	–	–[b]	–	–	–	650
pdBPV-1(142–6)	BPV[c]	–	–[b]	–	–	–	651
pTIIBPV(–), pTIIBPV(+)	BPV[a]	+[d]	mt-II$_A$[e]	mt-II$_A$	mt-II$_A$	mt-II$_A$	642
pdBPV-MMTneo (342–12)	BPV[c]	–	neo[f]	mt-I[g]	SV40t[h]	SV40e[i]	653
pAG0	–	–	tk[j]	tk	–	tk	605
pAG60	–	+	neo	tk	–	tk	664
pMK	–	+	tk	mt-I	–	tk	635
pASD11	SV40	+	$dhfr$[k]	ad[l]	ad/Ig[m]	$dhfr$[n]	655
pC6	SV40	+	tk	tk	–	tk	665

pSV0d	SV40^o	—	—	—	βG′	666
pHG, pMCG	SV40^o	+	bact $dhfr^p$	SV40e^q	βG	667

[a] Bovine papilloma virus, 69% fragment capable of *in vitro* transformation (see Section 5.2.3).

[b] No biochemical selectable marker, but possible to select for BPV-transformed phenotype in cells subject to density-dependent growth inhibition.

[c] Bovine papilloma virus, complete genome.

[d] Present but effect overcome by presence of *mt-II*$_A$ gene.

[e] Human metallothionein II$_A$ (intact genomic transcription unit, promoter regulated by Cd^{2+}).

[f] *E. coli* aminoglycoside 3′ phosphotransferase gene.

[g] Mouse metallothionein-I promoter, regulated by Cd^{2+}.

[h] SV40 t antigen mRNA splice sites.

[i] SV40 early polyadenylation site.

[j] From herpes simplex virus thymidine kinase transcription unit.

[k] Dihydrofolate reductase cDNA.

[l] Adenovirus major late promoter.

[m] Adenovirus major late transcription unit 5′ splice site + immunoglobulin V gene 3′ splice site.

[n] *dhfr* cDNA includes polyadenylation site.

[o] Origin only, therefore replication requires helper virus or COS cells.

[p] Bacterial gene for dihydrofolate reductase resistant to methotrexate.

[q] SV40 early promoter.

[r] From rabbit β-globin transcription unit.

TABLE 5.3. COSMID VECTORS USEFUL FOR INTRODUCTION OF DNA INTO MAMMALIAN CELLS

Vector	Eukaryotic Replication System	Selectable Marker	Processing Signals for Selectable Marker			References
			Promoter	Splice Site	Poly-adenylation Site	
cH1	—	$dhfr^a$	dhfr	dhfr	dhfr	656, 657
cCAD1, cCAD6	—	cad^b	cad	cad	cad	612
Homer III	$SV40^c$	—	—	—	—	668
pHEP	—	tk^d	tk	—	tk	658
pSAE	$SV40^c$	tk	tk	—	tk	658
pTM	—	neo^e	tk	—	tk	658
pMCS	$SV40^c$	neo	tk	—	tk	658
pTCF	—	neo	tk	—	tk	658
pGNC	—	tk	tk	—	tk	658
pNNL	—	gpt^f	$SV40e^g$	$SV40t^h$	SV40e	658
pTBE	—	—	—	—	—	658
pCGBPV7, pCGBPV9	BPV^i	neo	tk	—	tk	669

[a] Genomic Chinese hamster dihydrofolate reductase (intact transcription unit).
[b] Genomic Syrian hamster *cad* transcription unit (codes for CAD protein carrying first three enzyme activities of pyrimidine biosynthetic pathway).
[c] Origin only, therefore requires helper virus or COS cells.
[d] From herpes simplex virus thymidine kinase transcription unit.
[e] *E. coli* aminoglycoside 3' phosphotransferase gene.
[f] *E. coli* xanthine guanine phosphoribosyl transferase gene.
[g] From SV40 early transcription unit.
[h] SV40 small t antigen mRNA splice sites.
[i] Bovine papilloma virus.

agated over distances of several kilobases. Possible suggested mechanisms are that enhancers may act as chromatin organizers, RNA polymerase entry points, sites of attachment to the nuclear matrix, or regulators of local DNA superhelicity. That they may play a role in regulating chromatin structure is suggested by their presence in regions of papovavirus genomes not packaged into nucleosomes in the cell (645). The enhancer elements of different viruses contain a similar core sequence, but otherwise their sequences vary considerably (646).

TABLE 5.4. RETROVIRAL VECTORS USEFUL FOR INTRODUCTION
OF DNA INTO MAMMALIAN CELLS

| Vector | Retroviral Replication System[a] | Selectable Marker | Processing Signals for Selectable Marker | | References |
			Promoter	Poly-adenylation Site	
pMLVTK[b]	MLV[c]	tk[d]	tk	tk	670
HaSVtk5[e]	HaSV[f]	tk	tk	tk	671
TK2ΔΔter(R)	SNV[g]	tk	tk	SNV	659
MLV-NeoI	MLV	neo[h]	SV40e[i]	—[j]	654, 672
pLPAL[b]	MSV[k]/MLV	hgprt[l]	MSV	hgprt + MLV	673
pLPL[b]	MSV/MLV	hgprt	MSV	MLV	674

[a] Helper virus required in each case.
[b] Denotes plasmid clone from which retroviral vector may be recovered by transfection and superinfection.
[c] Moloney murine leukemia virus.
[d] From herpes simplex virus thymidine kinase transcription unit.
[e] Denotes bacteriophage λ clone from which retroviral vector may be recovered by transfection and superinfection.
[f] Harvey murine sarcoma virus.
[g] Spleen necrosis virus.
[h] E. coli aminoglycoside 3' phosphotransferase gene.
[i] SV40 early promoter.
[j] Polyadenylation site of neo transcription unit used in the construction was deleted by an in vivo recombination event: Whether or not the neo transcript is polyadenylated at another site has not been established.
[k] Moloney murine sarcoma virus.
[l] Human hypoxanthine guanine phosphoribosyl transferase cDNA.

Derivatives of these vectors with more than one inserted sequence have also been isolated, for example, pSV2-X-SVgpt (643) and pMDSG (647), in which the complete gpt-containing transcription unit of pSV2gpt is inserted into pSV2-X or into pMTVdhfr, respectively. Such vectors are useful for achieving chromosomal integration of genes for which no selective system exists.

The vector pCD-X has been developed to allow both the primary cloning of cDNA and its subsequent expression in animal cells (648). For primary cloning the vector is manipulated by restriction enzyme cutting and the addition of an oligo-dT tail to which mRNA is annealed via its polyA sequence. The cDNA is then synthesized *in situ*, the oligo-dT sequence functioning as a primer for reverse transcriptase (649). Following second strand synthesis and recyclization, the double-stranded cDNA sequence is situated within a transcription unit built up from the SV40 early promoter and late splicing and polyadenylation sequences. Alternative splicing sites are provided in case the cDNA sequences corresponding to the 5' end of the mRNA coding sequence are missing: If splicing occurs at the 16s sites, the first AUG triplet in the resulting RNA occurs in the sequence specified by the cDNA, while splicing at the 19s sites results in an RNA that has an AUG in its vector-specified region, and is thus translated as a fused protein. The splice sites are placed between the promoter and the coding sequence so that splicing can still occur even if there is a polyadenylation sequence within the cDNA.

The vectors listed in Tables 5.2, 5.3, and 5.4, with the exception of pTBE (Table 5.3), are capable of replication in mammalian cells or carry a selectable marker expressed in mammalian cells, or satisfy both these conditions. Those listed in Table 5.2 all carry pBR322-derived plasmid sequences, although some are constructed from derivatives of pBR322 lacking the "poison sequences." The vectors containing BPV sequences undergo episomal replication in rodent cells without integration, yielding stably transformed permanent cell lines carrying between 10 and 100 copies of the episomal recombinant genome (Section 5.2.3). This coupled with their ability to replicate episomally in *E. coli*, where they can be selected by virtue of the ampicillin-resistance marker derived from pBR322, enables them to be used as shuttle vectors. As indicated in Table 5.2, some carry the entire BPV genome, while others carry only the 69% fragment capable of *in vitro* transformation (Section 5.2.3). Facilitating sequences required for efficient transformation by vectors containing the 69% fragment are present in the remainder of the genome, but it appears that the β-globin gene insert present in pBPVβ1 and pBPVβ3 can fulfill a similar function, as can certain other eukaryotic sequences (642,650–652). pTIIBPV(–) and pTIIBPV(+) yield transformants expressing high levels of human metallothionein II_A which can be selected on the basis of increased resistance to cadmium salts (642). Insertion of selectable markers into BPV-based vectors can influence their stability. The vector pdBPV-MMT*neo* (342-12), which carries the *neo* marker, is stable, while vectors carrying the Eco*gpt* and HSV*tk* markers have undergone genomic rearrangement, concatemerization, and/or integration under selective pressure (653, and references cited therein). The herpes simplex virus *tk* transcription unit has been exploited in many of the vectors of Tables 5.2, 5.3, and 5.4: It contains no introns, but functions adequately to ensure the expression of the *neo* marker (623,654). Plasmid pASD11 contains a complete SV40 genome: The corresponding plasmid without SV40 sequences is incapable of transfecting DHFR⁻ CHO cells to DHFR⁺ (655). The increased transfection efficiency of pASD11 is probably due to the presence of the SV40 enhancer sequences (see above).

The vectors listed in Table 5.3 are cosmid vectors, that is, they are capable of replication in *E. coli* as plasmids but in addition carry the ligated cohesive end (*cos*) sites of bacteriophage λ to facilitate the cloning of long stretches of genomic DNA (see ref. 619). The entire genomic transcription units of dihydrofolate reductase and CAD protein have been cloned in cosmids and can be used as selective markers for transfer to mammalian cells by protoplast fusion (612,656,657). All the vectors listed in Table 5.3 which carry eukaryotic replication origins are constructed from derivatives of pBR322 from which the poison sequences have been deleted to facilitate replication in the appropriate mammalian host cells. Cosmids pHEP, pSAE, pTM, and pMCS carry compact selectable marker genes and are designed to accommodate, in addition, a long stretch of genomic DNA which does not itself carry a selectable gene (658). The derivatives pTCF, pGNC, pNNL, and pTBE have a repositioned cloning site and are designed to allow exchange of the selectable marker to which the inserted genomic DNA is linked (658). Vectors pCGBPV7 and pCGBPV9 are shuttle cosmids designed to allow episomal replication of long genomic sequences in both *E. coli* and rodent cells.

A number of retroviral vectors containing selectable markers have now been developed (Table 5.4): In constructing such vectors care must be taken that the polyadenylation site of the inserted gene does not interrupt the synthesis of the viral RNA genome (659). These vectors are well suited to applications where it is desired to achieve chromosomal integration of the foreign gene with the minimum of rearrangement (Section 5.2.3).

5.3. EXPRESSION OF GENES INTRODUCED INTO MAMMALIAN CELLS BY DNA-MEDIATED GENE TRANSFER

The factors that control the expression of genes introduced into mammalian cells are incompletely understood. While sequence rearrangements can account for some anomalies in gene expression (Section 5.2.2) they provide by no means a complete explanation. Some L cell transfectants expressing the HSV*tk* gene yield TK⁻ derivatives only at low frequency (of the order of 10^{-6}) while others do so at frequencies of 1% or higher. In the former group the loss of expression is associated with a change in the degree of methylation, while in the latter there is a change in sensitivity of the gene to DNAse I digestion without an accompanying change in methylation pattern, and the effect extends to other genes introduced on the same vector over distances as large as 15–20 kb and irrespective of gene orientation, suggesting that the difference in expression is mediated by a difference in DNA conformation (674, and references cited therein). It is presumed that the difference in behavior of the two groups of transfectants is due to differences in the site of chromosomal integration.

Genes normally expressed in a tissue-specific manner may be expressed when transfected into inappropriate cell types (675–677). Aberrant tissue-specific regulation of expression may also be seen in transgenic mice (Section 6.2). However, some success has been achieved recently in obtaining tissue-specific regulation of introduced genes. Chao et al. (678) found that in mouse erythroleukemia cells transfected with either a human β-globin gene or a fusion product constructed from mouse and human β-globin genes the level of transcription of the introduced gene was elevated 5 to 50-fold when differentiation was induced with agents such as DMSO. No corresponding elevation was seen with the *aprt* gene in these cells, or with globin genes introduced into fibroblasts. A number of groups have reported that transfection of immunoglobulin genes, isolated originally from myeloma cells, into lymphoid and nonlymphoid recipient lines leads to expression in the former case only (679–682). In one study the *neo* gene was present on the same plasmid and was expressed at comparable levels in all the recipient cell lines (680). Both in light-chain and in heavy-chain genes the sequences responsible for this tissue-specific expression have been localized to a short region within the intron between the J and C gene segments (679,681). Using a heavy-chain gene it has been shown that the effect of this region is still exerted either if it is isolated and reinserted in the opposite direction, or if it is translocated upstream of the promoter (681). In this respect it resembles a viral enhancer sequence (Sections 5.2.4 and 5.2.6), and indeed it contains a sequence homologous to the conserved "core" region of viral enhancers (681). If inserted into a plasmid together with a β-globin gene and the SV40 T antigen gene it permits expression of these genes after transfection into myeloma, but not nonlymphoid, recipient cells (683). The identification of an enhancer in the J-C intron provides an explanation for the activation of transcription of V region genes which occurs when they are translocated to the site of the J region in the course of lymphocyte differentiation (684). That enhancers may play a more general role in the control of tissue-specific gene expression is suggested by the observations of Walker et al. (167). They isolated upstream fragments several hundred nucleotides long from rat and human insulin genes and from the rat chymotrypsin B gene, inserted them into a plasmid in a position upstream of the *cat* transcription unit from pSV0*cat* (Table 5.1), and examined the levels of CAT activity after transfection of the following cell lines: (a) HIT (Table 1.5), which synthesizes insulin; (b) AR4-2J (Table 1.9), which synthesizes chymotrypsin mRNA; (c) CHO, which synthesizes neither. The overall efficiency of these cell lines as recipients in transfection differed considerably, but this effect could be eliminated by normalizing all results to those obtained in the same cell line using the Rous sarcoma virus promoter to activate the expression of CAT. On this basis, the upstream sequences from the insulin genes were 50- to 200-fold more active in HIT cells than in the other two lines, and those from the chymotrypsin gene were most active in AR4-2J cells by a similar margin. Deletion analysis localized the 5' border of the active sequence between 168 and 302 nucleotides from the transcription start site in all three cases, and the implicated region of the rat insulin I gene contains a sequence homologous to a viral enhancer sequence. It has been suggested that the tissue specificity shown by cellular enhancer sequences may be due to regulation by tissue-specific binding proteins (685).

The enhancer sequences of the immunoglobulin gene is probably also responsible for the tissue-specific expression which has been achieved in transgenic mice (Section 6.2).

5.4. *IN VITRO* MUTAGENESIS OF DNA

A variety of techniques are now available whereby it is possible to introduce mutations at chosen sites or within chosen regions of a DNA segment cloned in a vector: Such mutations include large deletions or insertions, for example, of entire restriction enzyme fragments, smaller deletions or insertions of a few bases, and defined substitutions (reviewed in ref. 686). Introduction of such mutated DNA into mammalian cells provides a further route for the introduction of genetic markers which is now beginning to be exploited.

Bacterial and bacteriophage genetics owe much to the analysis of suppressible nonsense mutations in essential genes (687). As conditional lethals, such mutations have the substantial advantage that, in the absence of suppression, gene function is completely lacking, in contrast to temperature-sensitive mutations which are often "leaky." Laski et al. (688) produced an amber suppressor gene by *in vitro* mutagenesis of the anticodon of a *Xenopus laevis* tyrosine transfer RNA gene cloned in a single-strand bacteriophage vector. The suppressor gene was then recloned into the late transcription unit of an SV40 vector and introduced into CV-1 monkey cells. It could be shown to function in these cells by its ability to suppress a viral amber mutation and by the ability of tRNA extracted from the infected cells to function in an *in vitro* suppression assay. However, no permanent cell line expressing the suppressor gene was obtained by this means. This further objective was achieved by introduction of the gene into L cells into which genes carrying amber mutations had previously been introduced to allow selection for the suppressor function (634). In order to introduce amber mutation-carrying genes, the HSV*tk*, *gpt*, and *neo* genes (Section 5.2.5) were separately cloned into plasmid vectors where their expression was under the control of the ASV-LTR (Section 5.2.4) which functions as a promoter in both bacterial and mammalian cells. (However, in practice, expression of the HSV*tk* gene from the ASV-LTR in *E. coli* was not sufficiently strong to allow the subsequent manipulations, and a modified protocol had to be used in this case.) The vectors were then mutagenized *in vitro*, transfected into *E. coli*, and amber mutants isolated by standard techniques of bacterial genetics. The plasmids containing the amber mutations were then microinjected into L cells along with a further plasmid containing a different selectable marker: In this way it was possible to achieve integration of all three amber mutant genes together with the selected marker in a single operation. By introduction of the suppressor gene and selection for *tk*, *gpt*, or *neo* function, stable cell lines exhibiting suppression of all three amber mutations could then be obtained. Using a similar strategy an amber suppressor gene has been constructed from a human lysine tRNA gene (689) and one may anticipate that a range of nonsense suppressors inserting different amino acids will soon be available in mammalian cell lines, and that their impact on mammalian cell and viral genetics will be as great as in prokaryotic systems.

The scope of *in vitro* mutagenesis studies with cloned mammalian DNA would be greatly broadened, of course, if it were possible to replace one or both of the originally resident alleles of a cellular gene by a mutated copy of the same gene introduced into the cell. Such allele replacement can be achieved in yeast as a result of integration of a recombinant episome specifically at the homologous chromosomal locus followed by its excision (690). However, it has not yet been feasible with mammalian vectors, because those that undergo chromosomal integration do so at multiple sites. Whether this is a problem inherent in the greater genome complexity of mammalian cells, or whether it can be overcome by the use of appropriately engineered vectors or recipient cells, is a question to be answered by future research.

5.5. ONCOGENES

Without doubt the outstanding success achieved by DNA-mediated gene transfer into mammalian cells has been the demonstration of oncogenes in the DNA of certain tumors (691). An in-depth review of this rapidly expanding subject would be inappropriate here, and would certainly be out-of-date before publication. Therefore, only a brief sketch is given.

Oncogenes were originally defined as sequences present in strongly transforming retroviruses which conferred upon them the ability to induce tumors in an appropriate host, and genetic analysis of such viruses has identified more than a dozen oncogenes (691). That these genes are closely related to normal cellular genes has been known for some time but the key to further progress was the demonstration (692) that DNA from cell lines derived from certain chemically induced mouse tumors, when introduced into NIH-3T3 mouse cells by transfection, is capable of transforming them to a phenotype characterized by lack of density-dependent growth inhibition, anchorage-independence, and tumorigenicity in newborn mice, while normal cell DNA lacks this capability (see below, however). Further work established that whereas the use of NIH-3T3 cells as recipients for transfection appears to be crucial to the success of such experiments, the ability to transform them is not restricted to tumor DNA of mouse origin and in particular is demonstrable in the DNA of cell lines derived from certain human tumors (and, indeed, in DNA taken directly from some primary human tumors; ref. 693). The active sequences in human bladder and lung carcinoma DNAs were found to be homologous to the *ras* oncogenes of Harvey (*ras*^H) and Kirsten (*ras*^K) sarcoma viruses (691).

Further work has cast considerable light on the relationship between the protooncogene sequences present in normal cells and their counterparts in malignant cells. The oncogenes of the EJ and T24 bladder carcinoma cell lines and the Hs242 lung carcinoma cell line differ from the corresponding normal protooncogene by a single base change (694–696). However, this is by no means the only way in which protooncogenes can be activated: Transposition next to a strong promoter and gene amplification are also effective (697). Examples of the former are seen both in protooncogenes linked to viral LTRs by genetic manipulation and in the action of weakly transforming retroviruses, which do not themselves carry an oncogene but undergo insertion at low frequency adjacent to a protooncogene of the recipient cell, thereby bringing it under control of the viral promoter. A similar mechanism undoubtedly underlies the observation that characteristic chromosome translocations associated with certain tumors bring protooncogenes into close proximity with a region of a different chromosome: In at least one case, that of Burkitt's lymphoma, the latter region is one known to be actively transcribed in the affected tissue, namely, the immunoglobulin heavy-chain gene (698). It probably also accounts for the observation that DNA from normal cells is active at low efficiency in the NIH-3T3 assay if fragmented into roughly gene-size pieces, the resulting transformed cells being an efficient source of transforming DNA provided no such fragmentation is carried out: Presumably the concatenation that occurs following DNA uptake (Section 5.1) occasionally places a protooncogene adjacent to one of the cell's own strong promoters (699).

In spite of this multiplicity in activation mechanisms the only oncogenes with which it has been possible to achieve directly transformation of normal cells are those of the strongly transforming retroviruses which both carry a mutation and are inserted adjacent to the strong viral promoter (697). For other oncogenes the use of the NIH-3T3 permanent line as recipient is necessary. This has been interpreted in terms of the known requirement for multiple events in the development of a tumor from a normal cell (700): According to this interpretation all but one of these events occurred during the establishment of the NIH-3T3 line, which therefore provides an assay for the remaining event. Support for this idea has come from recent experiments in which normal rodent fibroblasts could be transformed by the *ras*^H gene from human bladder carcinoma-derived cell lines only if (a) cotransformation was performed either with a second, virally promoted oncogene (*myc*, 701) or with a DNA tumor virus gene (701,702), *or* (b) the cells were first immortalized with a physical or chemical carcinogen (703), *or* (c) the gene was in a very high-expression vector (704). Even in the case of co-transformation with *myc*, sarcomas with only limited growth capacity were obtained, suggesting the need for a third event to obtain tumors with unlimited growth potential (701). Cotransformation assays may allow the detection of transforming activity in the DNA of tumors not active in the NIH-3T3 assay. Another route by which this may be achieved is the development of alternative recipient cell lines to NIH-3T3 in which the residual event required for transformation is a different one.

An important clue to the mechanism of oncogene action is provided by the recent finding that the *sis* oncogene of Simian sarcoma virus encodes a product with strong sequence homology to human platelet-derived growth factor (705,706). This, together with the observation that the cellular homologue of *sis* is activated in tumors derived from cell types sensitive to PDGF, suggests that the gene product of *sis* may function by completing a growth-stimulating positive feedback loop (autocrine stimulation; ref. 707). The protein products of some other oncogenes have a tyrosine-specific protein kinase activity reminiscent of that shown by receptors for growth factors on binding the cognate growth factor, suggesting that they may function likewise by completing a positive feedback loop (707).

Thus, at present, research into oncogenes stands at a very exciting point. It has provided a satisfying theoretical framework within which the actions of physical, chemical, and viral carcinogens are readily accommodated. Although as yet its relevance has been demonstrated only to a minority of tumors, and even there only to individual steps of a multistep process, there is no doubt that it will provide a springboard for further advance.

6

CELL GENETICS AND THE GERM LINE

The techniques described earlier in this book allow the somatic cell geneticist to obtain information which complements that available by classical genetic means. It has become feasible recently to use these techniques to introduce genetic markers into the germ line of experimental animals. In this way the somatic cell genetic approach may be applied to the study of the developing embryo and of the intact adult animal, and may also be combined with classical genetic analysis of the same system. Two routes are available for introducing genetic markers into the germ line by somatic cell genetic means. One is to use cell lines derived from teratocarcinomas or from normal early embryos, which can be cultured and then returned to the environment of the early embryo, where they can undergo normal development and contribute to the tissues of a chimeric animal including its germ cells (Section 6.1). The other is to carry out manipulations such as DNA-mediated gene transfer on the egg itself and then allow it to resume its normal development (Section 6.2).

6.1. TERATOCARCINOMAS

Teratocarcinomas are tumors occurring in a variety of mammals and birds, most commonly in the gonads but also in extragonadal sites. Their characteristic feature is that in histological section they show a chaotic patchwork appearance due to the presence of areas of a range of differentiated tissues which commonly include derivatives of all three embryonic germ layers (see refs. 708–712 for reviews). This is due to the presence of a developmentally pluripotent malignant cell population, the embryonal carcinoma (ec) cells, which give rise to the other tissues present by differentiation within the tumor. The differentiated tissues once formed are benign and the malignancy of the tumor is due entirely to the residual embryonal carcinoma cells it contains: If all the embryonal carcinoma cells in a particular tumor differentiate, the result is a benign tumor or teratoma. An alternative terminology designates the benign teratoma TD (teratoma, differentiated) and subdivides teratocarcinomas into MTU and MTI (malignant teratoma, undifferentiated and intermediate, respectively). In what follows the unqualified term "teratoma" is used to embrace both benign and malignant tumors, and "teratocarcinoma" to refer specifically to malignant tumors.

Two inbred strains of mice, 129 and LT, have a high spontaneous frequency of teratomas, which are testicular in strain 129 and ovarian in strain LT.

For this reason the mouse has become the preferred experimental animal for teratoma research. In addition, it is possible to induce teratomas by grafting either the genital ridges of male embryos of strain 129, or early embryos of most strains, to an ectopic site in a histocompatible adult. It has been established that the spontaneous testicular teratomas of strain 129 and those produced by genital ridge grafts derive from primordial germ cells, whereas those produced by early embryo grafts derive from the epiblast cells (often misleadingly designated embryonic ectoderm), the cells that give rise to all three germ layers of the embryo proper. The ovarian teratomas of strain LT arise because the oocytes of this strain undergo spontaneous parthenogenesis within the ovary and develop to a stage corresponding to an early postimplantation embryo, after which a high proportion give rise to teratomas. Presumably in these spontaneous ectopic embryos, as with the grafts, it is the epiblast cells that are responsible. The pathways by which teratomas arise from epiblast cells and from primordial germ cells may not be independent: Although it can be excluded that teratogenesis from epiblast cells involves primordial germ cells as an intermediate stage, it cannot be excluded that epiblast cells are intermediates in teratogenesis from primordial germ cells (710). The strain specificity of teratogenesis from primordial germ cells makes it reasonable to postulate that a genetic lesion is responsible, but the readiness with which epiblast cells from most strains are converted to teratomas in the absence of any obvious oncogenic stimulus cannot be explained in this way. Indeed, the evidence is now strong that embryonal carcinoma cells do not differ intrinsically from normal epiblast cells, but are malignant because of an abnormal microenvironment. When introduced into the environment of an early embryo (Section 6.1.3) they will contribute to the development of the normal tissues of a chimeric mouse. As with any other tumor cell, however, growth within the primary tumor, in transplantable tumor lines and in tissue culture imposes selection for faster-growing variant cells that do differ in their properties from normal epiblast cells, and, therefore, it is not surprising that some embryonal carcinoma cell lines do give rise to chimeras that develop tumors (Section 6.1.3). The fact that it is possible to obtain cell lines that contribute to completely normal development makes it reasonable to conclude that at the time of their origin, embryonal carcinoma cells differ from epiblast cells only in their microenvironment.

Generalization of these conclusions to species other than the mouse, however, is subject to some reservations. Attempts to produce teratocarcinomas by embryo grafts in other rodents have not been successful, benign teratomas only being obtained (708), and differences have been noted in the tissue composition of mouse and human teratocarcinomas, the latter commonly containing extraembryonic tissues (713). Teratocarcinomas have been induced in the rat by the action of Moloney murine sarcoma virus on embryos *in utero* (714) but the relationship of this etiology to those described above is not yet clear. It is presumed to be related to the development of benign teratomas which occurs in rats, hamsters, and mice when the visceral yolk sac is displaced after fetectomy, and for which an origin from cells in the visceral yolk sac other than germ cells has been demonstrated (715, and references cited therein).

6.1.1. Embryonal Carcinoma Cell Lines

A substantial number of embryonal carcinoma cell lines have been derived from teratocarcinomas (see ref. 716 for a recent compilation). In the case of mouse tumors this has generally been carried out via the production of tumor lines by serial transplantation of primary tumors and the adaptation of these lines to ascites growth. This results in the production of "embryoid bodies," cell aggregates in which a core of embryonal carcinoma cells is surrounded by an outer layer of endodermal cells. Other differentiated cell types may also be present but in all cases embryoid bodies are considerably enriched in embryonal carcinoma cells relative to the solid tumor and are therefore better starting material for the establishment of embryonal carcinoma cell lines. Mouse embryonal carcinoma cell lines differ widely in their properties: While some lines will grow in pure culture others require feeder layers (compare Section 1.9), even for growth at high cell density, if they are to be maintained in the undifferentiated state. Some cell lines in the latter class can be induced to differentiate in culture by plating in the absence of feeder cells, when they form embryoid bodies similar to those seen in the ascites, and these embryoid bodies when allowed to form outgrowths on a substratum undergo further differentiation with the production of a range of differentiated tissues. The requirement for feeder cells to prevent differentiation was reported by Oshima (717) to be spared by β-mercaptoethanol. A subsequent study, however, found that medium conditioned by feeder cells was also necessary, indicating that the feeder cells act by producing molecules that pass to the ec cells via the extracellular fluid (718). Some feeder-dependent cell lines will undergo differentiation if injected subcutaneously into mice, or under appropriate conditions in culture: Very high or low cell density favors differentiation in some cell lines, and a number of chemical agents with differentiation-promoting activity are also known. Retinoic acid induces differentiation in most embryonal carcinoma cell lines, usually to endoderm-like cells (719) although the differentiated cell type produced depends upon the ec cell line (720), the concentration of retinoic acid (721), the intracellular cyclic AMP content (722), and whether the cells are in monolayer or in suspension aggregates (723). As with differentiation of feeder-dependent cell lines, retinoic acid-induced differentiation is inhibited both by feeder cells and by medium conditioned by them (718,724). Other agents reported to promote differentiation in some ec cell lines include retinol (725), arotinoids (726), hexamethylene bis-acetamide (727,728), N,N dimethylacetamide (728), polybrene (728), dimethyl sulfoxide (729), 5-azacytidine (730), and 5-bromodeoxyuridine (731). The range of differentiated tissues seen in tumors formed by injection of different cell lines into mice varies widely, but this appears to reflect a difference in overall ability to differentiate rather than commitment to different lineages. The overall ability to differentiate tends to decrease with increase in passage number, probably because of selection for faster-growing variants. Some cell lines have completely lost their ability to differentiate in tumors, although even these lines in general still respond to retinoic acid in culture.

Isolation of human embryonal carcinoma cell lines is more difficult because this must either be carried out directly from the primary tumor, where the proportion of embryonal carcinoma cells is usually very low, or from xenografts which are subject to more selective pressure than are grafts between mice of the same inbred strain. This probably explains why no human ec cell line has yet been isolated which shows a broad range of differentiation *in vitro* (732). It is probably significant that one of the most promising human ec cell lines, PA-1, was obtained from an ascitic metastasis (732).

6.1.2. Differentiated Cell Lines

Differentiation of embryonal carcinoma cells is accompanied by a dramatic decrease in cloning efficiency. Nevertheless, a number of teratocarcinoma-derived differentiated cell lines have been isolated either from tumor lines passaged *in vivo* or from embryonal carcinoma cell lines that have undergone differentiation *in vivo* or *in vitro* (733, and references cited therein). These lines provide an addition to the list of cell lines that can be used to study differentiated function (Section 1.7). Different clonal cell lines derived from the same clonal embryonal carcinoma line represent homogeneous populations of cells each expressing a subset of the genetic information present in a common ancestral genome, and provide a means of studying developmental commitment and the regulation of gene expression.

6.1.3. Contribution of Cells from Embryonal Carcinoma Cell Lines to Chimeric Mice

Chimeric mice have been obtained both by microinjecting embryonal carcinoma cells into blastocyst-stage embryos which are then allowed to implant in the uterus of a foster mother (reviewed in ref. 710) and by aggregating clumps of embryonal carcinoma cells with morula-stage embryos from which the zona pellucida has been removed, the aggregates being allowed to develop in culture to the blastocyst stage before implantation in a foster mother (734). Results obtained with different cell lines, however, differ widely and, in general, are less successful than with cells taken directly from the *in vivo* transplant lines (710). Some cell lines give very little contribution to the embryo proper (as distinct from the extraembryonic part of the conceptus). Others contribute to the somatic tissues of the embryo but not to its germ line (see ref. 735 for a recent compilation) and so far only one cell line (METT-1, ref. 736) has been reported to give germ-line chimerism (in two of ten females overtly chimeric for coat-color markers). When groups of cells from some ec lines are used the chimeric mice develop tumors with the histological appearance of poorly differentiated teratocarcinomas. Mintz and co-workers found that when single ec cells were used they contributed either to normal tissues or to tumors but not to both, suggesting that tumors arise from variant cells that resist the normalizing influence of the blastocyst environment. However, Rossant and McBurney (737) found a clear-cut case of contribution of a single injected cell both to normal tissue and to tumor, as well as several cases where a single cell gave rise to both normal tissue and tissue that was in some way abnormal. Regulation of the proliferation of ec cells by the blastocyst *in vitro* is largely complete within 24 hr and appears to operate at the G1 phase of the growth cycle (738). What determines the ability of ec cells to colonize the germ line of chimeric mice is not yet clear: While a visibly normal diploid karyotype is probably a necessary criterion, it is clearly not a sufficient one (710). However, it is likely that *in vitro* culture conditions confer a selective advantage on cell variants with submicroscopic chromosomal lesions and that more success will be achieved as culture techniques improve.

A number of genetic markers have been used to distinguish between cells of host and tumor origin in chimeras (710). Until recently, however, there was no technique that allowed the identification of the origin of each cell in histological sections and could be applied to all tissues. This gap has now been filled in the case of chimeras between the murine species *M. musculus* and *M. caroli* by a technique that uses *in situ* nucleic acid hybridization to detect a repetitive DNA sequence present in *M. musculus* but not in *M. caroli* (739–741).

6.1.4. Derivation of Embryonal Carcinoma-Like Cell Lines from Normal Embryos

If indeed embryonal carcinoma cells are intrinsically normal epiblast cells rendered malignant by an abnormal microenvironment then one would predict that it should be possible to isolate embryonal carcinoma-like cell lines by direct explanation of normal embryos into culture. This was first achieved by Evans and Kaufman (742) who, in order to obtain preimplantation embryos that were as large as possible and thereby to maximize the number of cells available for culture, delayed the implantation of their embryos by artificially inducing a state of diapause. Martin (743) was able to dispense with the need to induce diapause by using medium conditioned by teratocarcinoma stem cells, and, indeed, it is now possible to obtain embryonal carcinoma-like cell lines without the use of either *in vivo* delay or conditioned medium (744,745). These cell lines are termed EK cells (742) or embryonal stem cells (743). This technique greatly facilitates the establishment of pluripotent cell lines from genetically marked mouse strains with the minimum of karyotypic abnormality, and has already been applied to the isolation of cell lines from embryos homozygous for the T-locus mutation t^{w5} (746), and from parthenogenetically activated embryos (747). All the cell lines obtained in the latter study had undergone diploidization, with the loss of part of one copy of the X chromosome (748): The reason for this is not yet known.

6.1.5.   Growth of Embryonal Carcinoma Cells in Chemically Defined Media

The requirements for the proliferation and differentiation of embryonal carcinoma cells have recently been reviewed (749). Serum-free growth of the embryonal carcinoma cell line F9 was first achieved by Rizzino and Sato (750) in a medium designated EM3 in which serum was replaced by fetuin (a crude serum glycoprotein preparation), insulin, transferrin, and β-mercaptoethanol. EM3 supports high-density growth and retinoic acid-induced differentiation, but not clonal growth. Subsequently, it was found that the requirement for fetuin and β-mercaptoethanol could be substituted by fibronectin (751) and a further improvement was noted if laminin was substituted for fibronectin (752,753). However, the embryonal carcinoma cell lines OC15S1 and 1003 do not proliferate in EM3 but differentiate and then die (754). Survival of the differentiated cells can be achieved by supplementation of EM3: Rizzino (754) reported that in EM3 supplemented with high-density lipoproteins endoderm-like cells were formed from cell line 1003, while Darmon and collaborators observed neural differentiation from the same cell line in EM3 supplemented with selenous acid (755–757). In the ec cell line PC13, which lacks high affinity insulin receptors, serum-free growth can be achieved in medium containing transferrin, high-density lipoprotein, and low-density lipoprotein (758). As the cells differentiate, receptors for growth factors such as insulin and epidermal growth factor appear, together with a growth requirement for such factors (759–762). Factors with transforming growth factor activity, to which the differentiated cells respond, are in fact produced by F9 and PC13 embryonal carcinoma cells (762).

6.1.6.   Variant Embryonal Carcinoma Cell Lines

Two novel approaches are possible using variant embryonal carcinoma cell lines isolated *in vitro*. First, by studying the developmental capacity of such variants one can obtain information about the developmental role of the altered function. Second, by colonizing the germ line of chimeric mice it should be possible to produce new phenotypes of mutant animals. The first approach does not require colonization of the germ line in the strict sense, since the differentiation of the variant embryonal carcinoma cells can be studied in culture, in tumors, and in the somatic tissues of chimeras.

It is possible to insert markers of general utility in somatic cell genetic manipulations, such as HGPRT⁻ and ouabain resistance, without interfering with developmental capacity (Table 6.1). It is also possible to isolate variants of intrinsic developmental interest. Variants which fail to respond to agents that induce differentiation are beginning to shed light on the mechanisms of action of such agents (763–766). Those selected for failure to differentiate in response to retinoic acid are also defective in their response to other differentiation stimuli, namely, aggregation or exposure to hexamethylene bis-acetamide (763,764) or DMSO, butyrate, or 6-thioguanine (765), while it has been possible to select variants which fail to differentiate in hexamethylene bis-acetamide (764) or DMSO (766) but which still respond to retinoic acid. Those failing to respond to DMSO are of particular interest: They were isolated from the ec line P19 which shows a concentration-dependent response when aggregates are exposed to retinoic acid, DMSO, butyric acid, or 6-thioguanine. Low concentrations of all these agents induce cardiac muscle differentiation and higher doses induce skeletal muscle differentiation. Still higher doses of retinoic acid induce differentiation to neurons and astrocytes, while corresponding concentrations of the other agents are toxic. The variant cell line showed no muscle differentiation in response to DMSO, butyrate, 6-thioguanine, or low concentrations of retinoic acid, but at high retinoic acid concentrations neuronal and glial differentiation occurred with the same dose-response as in the wild-type line (766). Metabolic cooperation-defective variants allow one to study the contributions of the gap junction to developmentally significant intercellular signaling (767–769). Such an approach, however, is complicated by the tendency for ec cell lines to lose capacity to differentiate with increase in passage number (Section 6.1.1), so that it is necessary to examine revertants to obtain good evidence for an association between a variant phenotype and altered developmental capacity. Reduced developmental capacity has been seen in several gap junction-defective variants (768,769) but in only one case has a revertant been studied, and since this failed to regain developmental capacity (768) there is no evidence as yet that the gap junction-defective phenotype is causally related to reduced developmental capacity.

Germ-line chimerism has not been obtained yet with a variant embryonal carcinoma cell line: All variants so far studied are derivatives of wild-type lines for which germ line chimerism has not been shown. The properties of variants isolated from cell lines such as METT-1 (Section 6.1.3) will be of great interest. Colonization of the germ line by an HGPRT⁻ variant

TABLE 6.1. VARIANT EMBRYONAL CARCINOMA CELL LINES

Variant Phenotype	Remarks	References
HGPRT[-]	Retention of ability to differentiate and colonize somatic tissues of chimeras	734, 771–773
TK[-]	Retention of ability to differentiate and colonize somatic tissues of chimeras	710, 774, 775
AK[-]	Retention of ability to differentiate and colonize somatic tissues of chimeras	477, 710
APRT[-]		776
Ouabain-resistant		777
DRB[a]-resistant		477
Podophyllotoxin-resistant		477
Canavanine-resistant	Altered arginine transport	445
PNA[b]-resistant	Retention of tumorigenicity and ability to differentiate	778

WGA[c]-resistant		779
Aphidicolin-resistant	Possible mutator strain	780
AraC[d]-resistant		780
Metabolic cooperation-deficient	Reduced gap junction incidence Reduced developmental capacity (see text)	767–769
Nontumorigenic	Immunogenic variant that confers protection against wild-type tumor	781, 782
Failure to differentiate in retinoic acid	Poor response also to induction of differentiation by other stimuli Some variants lack cRABP[e]	763–765
Differentiation in retinoic acid but not HMBA[f]	c-RABP-positive	764
Failure to differentiate in DMSO[g]	No muscle differentiation with DMSO or other agents but normal neural differentiation in retinoic acid (see text)	766

[a] 5,6 Dichloro 1-β-D-ribofuranosyl benzimidazole.
[b] Peanut agglutinin.
[c] Wheat germ agglutinin.
[d] Cytosine arabinoside.
[e] Cytoplasmic retinoic-acid-binding protein.
[f] Hexamethylene bis-acetamide.
[g] Dimethyl sulfoxide.

would provide the first available model in an experimental animal of the human inborn error Lesch–Nyhan Syndrome (770) and the ability to achieve germ-line chimerism routinely for other variants would make available a very powerful system for genetic analysis in which the approaches of somatic cell genetics and classical genetics could be combined.

6.1.7. Embryonal Carcinoma Hybrids, Cybrids, and Karyobrids

Most studies of embryonal carcinoma hybrids have been carried out on the products of intraspecies fusions because interspecies hybrids are difficult to isolate (783,784), although heterokaryons are readily obtained (783). The frequency at which even intraspecies hybrids are obtained is somewhat low compared with the average intraspecies fusion, and, since interspecies heterokaryons in general give rise to viable hybrids at a lower frequency than those of a comparable intraspecies fusion, it is difficult to say whether the problem is due to a species barrier or to generalized poor performance of embryonal carcinoma cells as fusion partners. Differentiated progeny of embryonal carcinoma cells readily form interspecies hybrids (783).

Early experiments showed that when embryonal carcinoma cells were fused to aneuploid cells of the established fibroblastoid line Cl-1D (an L cell derivative) the pluripotency of the embryonal carcinoma cells was extinguished and the hybrids formed tumors resembling fibrosarcomas (785). Subsequent experiments in which embryonal carcinoma cells have been fused to diploid thymocytes and near-diploid Friend erythroleukemia cells have demonstrated that, as with hybrids in general (Section 4.3.2), the properties of the hybrids are influenced by both genome dosage and growth habit. Anchorage-independent embryonal carcinoma × Friend cell hybrids are inducible for hemoglobin synthesis and show no embryonal carcinoma properties, while anchorage-dependent hybrids formed from the same two parents show the pluripotency of embryonal carcinoma and cannot be induced to synthesize hemoglobin (see 708 for references). In the first class of hybrids there is also activation of the embryonal carcinoma cell-derived globin gene (786). Interspecies hybrids in which small numbers of human chromosomes have been introduced into mouse ec cells by either microcell fusion or conventional hybridization differentiate similarly to the ec cell parent, and differentiated cell lines have been isolated from them (787). While hybrids between embryonal carcinoma cells and either intact 3T3 fibroblastoid cells or 3T3 karyoplasts show extinction of embryonal carcinoma properties, no such extinction is seen in cybrids formed with cytoplasts of 3T3 cells or myoblasts (788,789).

Two studies of the properties of hybrids between pluripotent and nullipotent cell lines have led to opposite conclusions. Rosenstraus et al. (777) concluded that such hybrids were pluripotent, but did not explain how the selective system used to isolate their hybrids (wild-type $\times$ HGPRT$^-$OUA^R plated in hypoxanthine + azaserine + ouabain on feeder layers of ouabain-resistant STO cells) eliminated the wild-type parent, which should be efficiently rescued from ouabain toxicity by metabolic cooperation with the feeder cells (204). In contrast, Oshima et al. (790) concluded that such hybrids were nullipotent but two flaws were present in their analysis. First, they used a selective system that allowed the growth only of those hybrids which, like the nullipotent parent, were feeder independent, the pluripotent parent being feeder dependent. Second, they did not carry out any karyotypic characterization, other than total DNA content per cell, of the tumors whose histology was used to assess developmental potency, so that one could not exclude the possibility of chromosome segregation during the growth of the tumors. In spite of the query about their selective system the conclusion drawn by Rosenstraus and co-workers is almost certainly correct as it has now been shown unequivocally that nullipotent embryonal carcinoma cells can acquire pluripotency by fusion with normal thymocytes or lens epithelial cells (791,792).

The ability of two interspecies hybrids to colonize the tissues of chimeric mice has been studied. Hybrid cells carrying a single human chromosome 17 were shown to colonize a number of tissues but unequivocal evidence for the expression of human markers could not be obtained (774). A hybrid between mouse ec cells and rat hepatoma cells contributed preferentially to the liver and a limited number of other organs of endomesodermal origin. In this case the expression of a number of rat-specific isozymes was detectable: Some rat isozymes of restricted tissue distribution were not expressed in the hybrid cells in culture but were expressed *de novo* in the chimeric tissues, some in a tissue-specific manner (793,794).

Cybrids obtained by fusing ec cells to cytoplasts of melanoma cells selected for chloramphenicol resistance (compare Section 4.5) have been shown to colonize various tissues of chimeric mice. Although the cybrids stably retained chloramphenicol resistance in culture it was impossible for technical reasons to demonstrate this in the chimeric tissues and colonization was shown by examining markers coded by the nucleus (795). This demonstrates the feasibility of incorporating mitochondrial markers into mice via embryonal carcinoma cells: As pointed out in ref. 710, a female cell line would probably be required to obtain germ-line progeny because mitochondrial markers are likely to show maternal inheritance.

6.1.8. Molecular Somatic Cell Genetics of Embryonal Carcinoma Cells

Most of the vectors developed for the introduction of DNA into mammalian cells rely heavily on the use of processing signals from viral transcription units to achieve expression of inserted genes and/or selectable markers (Section 5.2.4). Their application to embryonal carcinoma cells is complicated by the fact that in these cells the growth of papovaviruses and retroviruses (as well as parvoviruses and the herpesvirus MVMV) is blocked (796,797). The block is relieved when the cells differentiate, but information regarding its nature is somewhat fragmentary. In SV40-infected embryonal carcinoma cells T antigen is not produced. Segal et al. (798) reported that the T antigen gene is transcribed but that the transcript does not undergo splicing, whereas Linnenbach et al. (799) found normal splicing of the transcript from the SV40 T-antigen gene carried in the pC6 vector integrated into embryonal carcinoma chromosome DNA. The reason for this discrepancy has yet to be resolved.

Polyoma virus fails to grow in F9 embryonal carcinoma cells at any temperature, but in embryonal carcinoma cells of the PCC4 line its growth is temperature sensitive (800). Both in PCC4 cells at 37°C and in F9 cells early mRNA is correctly spliced but is present at very low levels (801,802). Two classes of mutant polyoma virus have been isolated with altered growth properties in embryonal carcinoma cells: One class replicates in PCC4 cells at both 31° and 37°C but not in F9 cells, while the other replicates in both F9 and PCC4 cells. Both classes of mutant carry complex lesions that are all localized in a region of the polyoma genome which has been shown to function as an enhancer of transcription of the rabbit β_1-globin gene (reviewed in ref. 803). Wild-type polyoma also fails to grow in trophoblast cells: Mutants selected for growth in F9 cells also grow in trophoblast cells, but it is possible to select mutants that grow in trophoblast cells but not in F9. Mutants of the latter class also carry complex alterations in the same region. These results are of interest in relation to recent indications that enhancer sequences are important in achieving correct tissue-specific regulation of gene expression in transgenic mice (Section 6.2).

In the case of the retrovirus Moloney murine leukemia virus, the provirus undergoes *de novo* methylation, although with different embryonal carcinoma cell lines this has been reported to occur either immediately on integration (804) or with a delay of about 2 weeks (805,806). In the latter case no transcription occurred even before methylation, indicating that a further block to expression must exist.

A direct assay for the efficiency of various promoters in embryonal carcinoma cells has been developed by Herbomel et al. (807). They demonstrated that the SV40 early promoter, which in 3T6 cells was more efficient than either SV40 or polyoma late promoters, fell in F9 embryonal carcinoma cells to the same efficiency as that of the late promoters. A cellular promoter, that from the chicken collagen α2 type I gene, functioned weakly in both 3T6 and F9 cells, but in both cell types its efficiency could be increased by inserting the wild-type polyoma enhancer sequence 3' to the assay gene (chloramphenicol acetyl transferase). A small effect of the wild-type polyoma enhancer sequence was seen on the level of expression of the HSV*tk* gene in F9TK⁻ cells, but a far more dramatic increase in the level of expression was seen with the corresponding sequence from mutants selected for growth in F9 cells (808). This sequence functioned as an enhancer of HSV*tk* expression when inserted in either orientation upstream of the gene, but in only one direction downstream. If the enhancer/HSV*tk* construct was cotransfected with a vector carrying the enhancer alone the effect was reduced, suggesting that the enhancer functions by binding to a cell component that can be saturated. Unfortunately, this work was carried out with mycoplasma-infected cells, and although an attempt was made to decontaminate them by animal passage before use, the authors imply that this was not wholly successful (808). As yet few vectors have been successfully used to introduce DNA into embryonal carcinoma cells. The vector pC6 (665) allows expression of the HSV*tk* gene at sufficient levels for selection in F9TK⁻ cells, probably due to the fortunate combination of the HSV*tk* promoter and SV40 enhancer sequences in the same plasmid. HSV*tk* has been used as selective marker for the introduction of the human β-globin gene into embryonal carcinoma cells by cotransfection (775). Attempts to use the Berg vectors of Table 5.1, carrying selectable markers transcribed from the SV40 early promoter, in order to achieve selection in cells not carrying the TK⁻ marker, led to conflicting reports regarding their applicability to embryonal carcinoma cells. The reason for the discrepancies in the results obtained has to some extent been elucidated by the work of Nicolas and Berg (641). They were unable to select transfectant ec cells using pSV2-*neo* or pSV2-*gpt* (Table 5.1), but were able to construct vectors pSV*tk-gpt* and pSV*tk-neo*β (Table 5.1) in which the HSV*tk* promoter was inserted between the SV40 early promoter and the selectable marker gene, allowing transcription of the latter from the HSV*tk* promoter augmented by the SV40 enhancer sequences. These latter vectors gave transfectants with ec cells at high frequency. Furthermore, pSV*tk-neo*β could be used to introduce the pSV3-*gpt* genome by cotransfection. In general, such cotransfectants did not express either the Eco*gpt* gene or the SV40 T antigen gene of pSV3-*gpt* at appreciable levels from their respective copies of the SV40 early promoter, but, by selection in mycophenolic acid plus guanine, variant cells could be obtained which expressed both Eco*gpt* and T antigen. Such variant ec cells, presumably differing from wild-type cells in their interaction with the SV40 early promoter, could account for the rare successful transfections obtained in earlier work. However, unexplained findings remain: In cotransfectants carrying both pSV*tk-neo*β and pSV*tk-gpt* integrated into chromosomal DNA, both selectable markers continue to be expressed as long as selective growth conditions are maintained for either marker. In the absence of selection the expression of both markers is turned off without any gross change in genomic structure. Reapplication of selective conditions then selects for variants expressing both markers (641). Jami et al. (809) found that if F9TK⁻ cells were transfected with the plasmid pBB1 which carries both the *gpt*-containing transcription unit from pSV2-*gpt* and a separate HSV*tk* transcription unit, all transfectants selected for XGPRT activity also expressed TK but not vice versa. This supports the concept that expression of the *gpt* gene in ec cells requires a further event in addition to integration. However, in contrast to Nicolas and Berg (641), they reported that selection for *gpt* expression did not result in cells capable of supporting T antigen expression. They also constructed a recombinant plasmid pQS20 similar, but not identical, to pSV*tk-gpt*, and found that it was no more efficient in transfecting ec cells than was pSV2-*gpt*. The precise reason for the difference in the behavior of pQS20 and pSV*tk-gpt* remains to be elucidated.

Preliminary characterization of vectors related to pC6 (see above) has been reported (810). Replacement of the SV40 genome of pC6 by the genome of either wild-type polyoma, a polyoma mutant selected for growth in F9 ec cells, or Kirsten murine sarcoma virus results in vectors capable of transfecting F9TK⁻ cells, but no quantitative study of transfectant yield or *tk* gene expression has been reported yet. The vectors pTIIBPV(+) and pTIIBPV(−) (Table 5.2) have also been reported to be capable of transfecting F9 cells (642). However, now one may expect the construction of further vectors tailor-made for efficient gene expression in embryonal carcinoma cells, in which enhancer sequences such as those from polyoma mutants may be expected to play a significant role.

6.2. GENETIC MANIPULATION OF THE MOUSE EGG

Microsurgery of the fertilized mouse egg has now been refined to the point where it is possible to carry out manipulations such as removal of one or both pronuclei and introduction of an exogenous nucleus before allowing implantation in a foster mother (see ref. 811 for a review and 812 for a recent technical improvement). The results obtained from such experiments are currently the source of some controversy. Illmensee and collaborators (811) have reported (a) the production of viable homozygous diploid animals by removal of one pronucleus and diploidizing the other with cytochalasin B; (b) the production of live-born animals by microinjecting nuclei from cells of the inner cell mass of the blastocyst or from the epiblast of the early postimplantation embryo into enucleated fertilized eggs, a technique that enables cloned mice to be produced; and (c) microinjection of nuclei from parthenogenetically activated embryos into enucleated fertilized eggs to produce viable mice. However, other laboratories have been unable to reproduce these results and their authenticity is at present the subject of an enquiry (813).

No such doubts, however, apply to experiments in which cloned DNA sequences have been introduced into mouse eggs to produce "transgenic" mice. This advance has been made possible by the high freuqency of incorporation which can be achieved by microinjection into the nucleus (Section 5.1), thus circumventing the problem that selective techniques cannot be applied to the egg. A number of laboratories have reported successful incorporation of various exogenous genes into both somatic tissues and the germ-line of viable offspring (reviewed in 811). As with microinjection into tissue-culture cells the introduced sequences, in general, are found stably integrated at one or more chromosomal loci as multiple tandem copies, the loci differing in different animals. However, less success has been achieved with expression of the integrated genes: If expression occurs at all it is frequently in inappropriate tissues. Lacy et al. (814) have recently analyzed five lines of mice transgenic for DNA containing the rabbit β-globin gene. Each line was shown by *in situ* hybridization of probe DNA to metaphase chromosomes to have one or two loci of integration, each with between 3 and 40 tandem copies of the gene, the pattern of integration showing normal Mendelian inheritance. In none of the five lines was the gene expressed in erythroid cells, but in one line a low level of transcription was found in muscle and, in another line, in testis, these aberrant patterns of expression again being heritable traits. The results suggest that chromosomal location is dominant over any cis-acting control sequences brought in with the DNA. Nonetheless, some successes have been achieved in obtaining expression of the transgenome in appropriate tissues. Palmiter et al. (815) obtained seven mice transgenic for a DNA fragment containing the promoter of the mouse metallothionein-I gene fused to the structural gene of rat growth hormone, and six of them grew significantly larger than their littermates. Three of these mice had very high levels of the fusion mRNA in their liver and between 200- and 800-fold elevation of the serum growth hormone concentration. Zinc sulfate was included in the diet of the mice as an inducer of the metallothionein-I promoter but it is not clear whether this had any influence on expression. While the liver is not the normal site of synthesis of growth hormone it is the normal site of action of the metallothionein-I promoter: Whether expression in the transgenic mice occurs only in the liver, or whether it occurs also in other tissues, has not yet been reported. McKnight et al. (816) found that the chicken transferrin gene was preferentially expressed in the liver of transgenic mice, but still at a much lower level than in chicken liver. The most informative experiments, however, are those recently reported by Brinster et al. (817), who produced transgenic mice carrying the immunoglobulin κ gene from the myeloma MOPC-21. In first generation offspring from three different transgenic fathers, in which the chromosomal localization of the foreign sequences are presumably different from each other, expression was consistently seen in the spleen (presumably in B lymphocytes) and not in the liver. As such genes have been shown to contain enhancer sequences which allow expression after transfection into lymphoid but not nonlymphoid cell lines (Section 5.3), it seems likely that in transgenic mice this enhancer can override chromosomal position effects. Study of mice transgenic for κ genes carrying specific deletions should enable this hypothesis to be tested directly.

6.3. COMPARISON OF TECHNIQUES FOR INTRODUCING GENETIC MARKERS INTO THE GERM LINE

It is clear that microinjection into the mouse egg is the method of choice for the introduction of defined exogenous genes into the germ line of mice. Even when problems in obtaining germ-line chimerism with cell lines of embryonal carcinoma or embryonal stem cells are overcome, as it seems likely they will be, introduction of exogenous genes into mice via such cell lines will still be a more indirect and laborious route than introduction directly into the egg. Where these cell lines will continue to hold an advantage, however, is in the range of genetic techniques which can be applied to them, largely as a consequence of the possibility of using selective isolation techniques. It seems likely that both approaches will have a considerable role to play in furthering our understanding of the multicellular organism at the molecular level.

REFERENCES

1. T. T. Puck (1972). *The mammalian cell as a microorganism: Genetic and biochemical studies in vitro* (Holden-Day, San Francisco).

2. J. Littlefield (1976). *Variation, senescence and neoplasia in cultured somatic cells* (Harvard University Press, Cambridge, Mass.).

3. J. Morrow (1983). *Eukaryotic cell genetics* (John Wiley & Sons, New York).

4. G. Martin et al. (in press). *Somatic cell genetics* (John Wiley & Sons, New York).

5. C. T. Caskey and D. C. Robbins, eds. (1982). *Somatic cell genetics* (Plenum Press, New York).

6. J. W. Shay, ed. (1982). *Techniques in somatic cell genetics* (Plenum Press, New York).

7. J. Paul (1975). *Cell and tissue culture*, 5th ed. (Livingstone, Edinburgh).

8. R. I. Freshney (1983). *Culture of animal cells: A manual of basic technique* (Alan R. Liss/Wiley, New York).

9. P. F. Kruse, Jr. and M. K. Patterson, Jr., eds. (1973). *Tissue culture: Methods and applications* (Academic Press, New York).

10. W. B. Jacoby and I. H. Pastan, eds. (1979). *Cell culture*. Methods in Enzymology (S. P. Colowick and N. O. Kaplan, eds., vol. 58 (Academic Press, New York).

11. C. Waymouth, R. G. Ham, and P. J. Chapple, eds. (1981). *The growth requirements of vertebrate cells in vitro*. (Cambridge University Press, Cambridge, UK).

12. C. Waymouth (1981). In ref. 11, p. 33.

13. G. Sato, ed. (1981). *Functionally differentiated cell lines* (Alan R. Liss, New York).

14. P. Ralph (1981). In ref. 13, p. 117.

15. R. G. Ham and W. L. McKeehan (1979). In ref. 10, p. 44.

16. H. Eagle (1959). *Science* **130**: 432.

17. R. G. Ham (1965). *Proc. Natl. Acad. Sci. USA* **53**: 288.

18. W. L. McKeehan, K. A. McKeehan, and R. G. Ham (1981). In ref. 11, p. 223.

19. C. Waymouth (1981). In ref. 11, p. 105.

20. A. Leibovitz (1963). *Am. J. Hyg.* **78**: 173.

21. C. T. Ling, G. O. Gey, and V. Richters (1968). *Exp. Cell Res.* **52**: 469.

22. W. G. Taylor (1981). In ref. 11, p. 94.

23. A. Richter, K. K. Sanford, and V. J. Evans (1972). *J. Natl. Cancer Inst.* **49**: 1705.

24. J. R. Smith (1981). In ref. 11, p. 343.

25. M. E. King and A. A. Spector (1981). In ref. 11, p. 293.

26. R. G. Ham (1964). *Biochem. Biophys. Res. Commun.* **14**: 34.

27. J. I. Toohey (1975). *Proc. Natl. Acad. Sci. USA* **72**: 73.

28. R. Oshima (1978). *Differentiation* **11**: 149.

29. J. I. Toohey and M. J. Cline (1976). *Biochem. Biophys. Res. Commun.* **70**: 1275.

30. J. I. Toohey (1977). *Biochem. Biophys. Res. Commun.* **78**: 1273.

31. J. I. Toohey (1978). *Biochem. Biophys. Res. Commun.* **83**: 27.

32. J. I. Toohey (1982). *Biochem. Biophys. Res. Commun.* **109**: 313.

33. F. H. Nielsen (1981). In ref. 11, p. 68.

34. E. J. Underwood (1977). *Trace elements in human animal nutrition*, 3rd ed. (Academic Press, New York).

35. J. T. Rotruck, A. L. Pope, H. E. Ganther, A. B. Swanson, D. G. Hafeman, and W. G. Hoekstra (1973). *Science* **179**: 588.

36. J. Fogh, ed. (1973). *Contamination in tissue culture* (Academic Press, New York).

37. B. E. Cham and B. R. Knowles (1976). *J. Lipid Res.* **17**: 176.

38. G. H. Rothblat, L. Y. Arbogast, L. Ouellette, and B. V. Howard (1976). *In Vitro* **12**: 554.

39. R. G. Ham (1974). *In Vitro* **10**: 119.

40. R. G. Ham (1981). In ref. 11, p. 1.

41. W. L. McKeehan, D. Genereux, and R. G. Ham (1978). *Biochem. Biophys. Res. Commun.* **80**: 1013.

42. I. Hayashi, J. Larner, and G. Sato (1978). *In Vitro* **14**: 24.

43. J. Bottenstein, I. Hayashi, S. Hutchings, H. Masui, J. Mather, D. B. McClure, S. Ohasa, A. Rizzino, G. Sato, G. Serrero, R. Wolfe, and R. Wu (1979). In ref. 10, p. 94.

44. N. N. Iscove and F. Melchers (1978). *J. Exp. Med.* **147**: 923.

45. Y. Honma, T. Kasutabe, J. Okabe, and M. Hozumi (1979). *Exp. Cell Res.* **124**: 421.

46. C. H. Uittenbogaart, Y. Cantor, and J. L. Fahey (1983). *In vitro* **19**: 69.

47. C. Rappaport, J. P. Poole, and H. P. Rappaport (1960). *Exp. Cell Res.* **20**: 465.

48. F. Grinnell (1978). *Int. Rev. Cytol.* **53**: 65.

49. M. Höök, K. Rubin, A. Oldberg, B. Öbrink, and A. Vaheri (1977). *Biochem. Biophys. Res. Commun.* **79**: 726.

50. F. Grinnell and M. K. Feld (1979). *Cell* **17**: 117.

51. R. L. Ehrmann and G. O. Gey (1956). *J. Natl. Cancer Inst.* **16**: 1375.

52. H. K. Kleinman, E. B. McGoodwin, S. I. Rennard, and G. R. Martin (1979). *Anal. Biochem.* **94**: 308.

53. K. Rubin, Å Oldberg, M. Höök, and B. Öbrink (1978). *Exp. Cell Res.* **117**: 165.

54. E. G. Bernstine, M. L. Hooper, S. Grandchamp, and B. Ephrussi (1973). *Proc. Natl. Acad. Sci. USA* **70**: 3899.

55. W. L. McKeehan, K. A. McKeehan, and R. G. Ham (1981). In ref. 11, p. 118.

56. W. L. McKeehan and R. G. Ham (1976). *J. Cell Biol.* **71**: 727.

57. A. L. van Wesel (1973). In ref. 9, p. 372.

58. M. Butler, T. Imamura, J. Thomas, and W. G. Thilly (1983). *J. Cell Sci.* **61**: 351.

59. C. R. Keese and I. Giaever (1983). *Science* **219**: 1448.

60. C. R. Keese and I. Giaever (1983). *Proc. Natl. Acad. Sci. USA* **80**: 5622.

61. M. M. Bashor (1979). In ref. 11, p. 119.

62. T. Takaoka, S. Yasumoto, and H. Katsuka (1975). *Jpn. J. Exp. Med.* **45**: 317.

63. C. B. Wigley (1975). *Differentiation* **4**: 25.

64. S. Federoff (1967). *Exp. Cell Res.* **44**: 642.

65. D. Yaffe (1968). *Proc. Natl. Acad. Sci. USA* **61**: 477.

66. F. C. Jensen, R. B. Gwatkin, and J. D. Biggers (1964). *Exp. Cell Res.* **34**: 440.

67. F. Gilbert and B. R. Migeon (1975). *Cell* **5**: 11.

68. G. D. Stoner, C. C. Harris, G. A. Myers, B. F. Trump, and R. D. Connor (1980). *In Vitro* **16**: 399.

69. M. M. Webber and D. Chaproniere-Rickenberg (1980). *Cell Biol. Int. Rep.* **4**: 185.

70. P. K. A. Jensen and A. J. Therkelsen (1982). *In Vitro* **18**: 867.

71. J. P. Mather and G. H. Sato (1979). *Exp. Cell Res.* **124**: 215.

72. F. S. Ambesi-Impiombata, L. A. M. Parks, and H. G. Coon (1980). *Proc. Natl. Acad. Sci. USA* **77**: 3455.

73. M. C. Tsao, B. J. Walthall, and R. G. Ham (1982). *J. Cell Physiol.* **110**: 219.

74. W. L. McKeehan, P. S. Adams, and M. P. Rosser (1982). *In Vitro* **18**: 87.

75. L. A. Culp and P. H. Black (1972). *Biochemistry* **11**: 2161.

76. H. G. Coon (1966). *Proc. Natl. Acad. Sci. USA* **55**: 66.

77. G. M. Hodges, D. C. Livingston, and L. M. Franks (1973). *J. Cell Sci.* **12**: 887.

78. L. Hayflick (1965). *Exp. Cell Res.* **37**: 614.

79. G. M. Martin (1977). *Am. J. Pathol.* **89**: 484.

80. D. Rohme (1981). *Proc. Natl. Acad. Sci. USA* **78**: 5009.

81. J. G. Rheinwald and H. Green (1975). *Cell* **6**: 331.

82. R. Holliday, L. I. Huschtscha, G. M. Tarrant, and T. B. L. Kirkwood (1977). *Science* **198**: 366.

83. J. Prothero and J. A. Gallant (1981). *Proc. Natl. Acad. Sci. USA* **78**: 333.

84. W. C. Topp, D. Lane, and R. Pollack (1981). In *DNA tumor viruses* (J. Tooze, ed.) (Cold Spring Harbor Laboratory, New York), p. 205.

85. L. I. Huschtscha and R. Holliday (1983). *J. Cell Sci.* **63**: 77.

86. G. J. Todaro and H. Green (1963). *J. Cell Biol.* **17**: 299.

87. W. R. Earle (1943). *J. Natl. Cancer Inst.* **4**: 165.

88. K. K. Sanford, W. R. Earle, and G. D. Likely (1948). *J. Natl. Cancer Inst.* **9**: 229.

89. R. J. Kuchler and D. J. Merchant (1956). *Proc. Soc. Exp. Biol. Med.* **92**: 803.

90. S. A. Aaronson and G. J. Todaro (1968). *J. Cell Physiol.* **72**: 141.

91. T. T. Puck, S. J. Cieciura, and A. Robinson (1958). *J. Exp. Med.* **108**: 945.

92. F-T. Kao and T. T. Puck (1968). *Proc. Natl. Acad. Sci. USA* **60**: 1275.

93. I. A. McPherson and M. G. P. Stoker (1962). *Virology* **16**: 147.

94. V. G. Dev and R. Tantravahi (1982). In ref. 6, p. 493.

95. B. Dutrillaux (1975). *Chromosoma* **52**: 261.

96. B. Dutrillaux and E. Viegas-Pequignot (1981). *Human Genet.* **57**: 93.

97. J. J. Yunis (1983). *Science* **221**: 227.

98. B. U. Zabel, S. L. Naylor, A. Y. Sakaguchi, G. I. Bell, and T. B. Shows (1983). *Proc. Natl. Acad. Sci. USA* **80**: 6932.

99. B. Ephrussi (1972). *Hybridization of somatic cells* (Princeton University Press, Princeton, N.J.).

100. N. R. Ringertz and R. E. Savage (1976). *Cell hybrids* (Academic Press, New York).

101. J. M. Clark and J. A. Pateman (1978). *Exp. Cell Res.* **114**: 317.

102. J. M. Clark and J. A. Pateman (1978). *Exp. Cell Res.* **114**: 47.

103. M. Steinberg and V. Defendi (1979). *Proc. Natl. Acad. Sci. USA* **76**: 861.

104. H. Holtzer, S. J. Tapscott, G. S. Bennett, M. Pacifici, and R. Payette (1982). In *Expression of differentiated function in cancer cells* (R. Revoltella, ed.) (Raven Press, New York).

105. V. K. Singh and D. van Alstyne (1978). *Brain Res.* **155**: 418.

106. H. P. Morris (1965). *Adv. Cancer Res.* **9**: 227.

107. S. E. Pfeiffer, E. Barbarese, and S. Bhat (1981). In ref. 13, p. 141.

108. V. Buonassisi, G. Sato, and A. I. Cohen (1962). *Proc. Natl. Acad. Sci. USA* **48**: 1184.

109. B. P. Schimmer (1981). In ref. 13, p. 61.

110. W. Rutter, R. Pictet, and P. Morris (1973). *Ann. Rev. Biochem.* **42**: 601.

111. J. Morrow (1983). In ref. 3, p. 157.

112. H. Green and O. Kehinde (1974). *Cell* **1**: 113.

113. C. N. Christian, P. G. Nelson, J. Peacock, and M. Nirenberg (1977). *Science* **196**: 995.

114. B. W. Kimes and B. L. Brandt (1976). *Exp. Cell Res.* **98**: 367.

115. B. W. Kimes and B. L. Brandt (1976). *Exp. Cell Res.* **98**: 349.

116. K. Bulloch, W. B. Stallcup, and M. Cohn (1977). *Brain Res.* **135**: 25.

117. C. Waymouth, H. W. Chen, and B. G. Wood (1971). *In Vitro* **6**: 371.

118. D. Rintoul, J. Colofiore, and J. Morrow (1973). *Exp. Cell Res.* **78**: 414.

119. J. A. McRoberts, M. Taub, and M. H. Saier, Jr. (1981). In ref. 13, p. 117.

120. J. P. Mather and F. Haour (1981). In ref. 13, p. 93.

121. K. Nagao, K. Yakoro, and S. A. Aaronson (1980). *Science* **212**: 333.

122. R. Weinstein, M. B. Stemerman, D. E. MacIntyre, H. N. Steinberg, and T. Maciag (1981). *Blood* **58**: 110.

123. D. C. Williams, C. G. Boder, R. E. Toomey, D. C. Paul, C. C. Hillman, Jr., K. L. King, R. M. van Frank, and C. C. Johnston Jr. (1980). *Calcif. Tiss. Internat.* **30**: 233.

124. H. Sudo, H. A. Kodama, Y. Amagai, S. Yamamoto, and S. Kasai (1983). *J. Cell Biol.* **96**: 191.

125. T. Log, K. S. S. Chang, and Y. C. Hsu (1981). *Int. J. Cancer* **27**: 365.

126. J. Y. Chou (1978). *Proc. Natl. Acad. Sci. USA* **75**: 1409.

127. J. Y. Chou and S. E. Schlegel-Haueter (1981). *J. Cell Biol.* **89**: 216.

128. S. P. Banks-Schlegel and P. M. Howley (1983). *J. Cell Biol.* **96**: 330.

129. G. J. Giotta and M. Cohn (1981). *J. Cell Physiol.* **107**: 219.

130. R. F. Santerre, R. A. Cook, R. M. D. Crisel, J. D. Sharp, R. J. Schmidt, D. C. Williams, and C. P. Wilson (1981). *Proc. Natl. Acad. Sci. USA* **78**: 4339.

131. N. Rosenberg, D. Baltimore, and C. D. Scher (1975). *Proc. Natl. Acad. Sci. USA* **72**: 1932.

132. C. Friend (1981). In ref. 13, p. 235.

133. J. S. Greenberger, P. B. Davisson, P. J. Gans, and W. C. Moloney (1979). *Blood* **53**: 987.

134. N. G. Testa, T. M. Dexter, D. Scott, and N. M. Teich (1980). *Brit. J. Cancer* **41**: 33.

135. T. M. Dexter, T. D. Allen, D. Scott, and N. M. Teich (1979). *Nature* **277**: 471.

136. F. C. Bancroft (1981). In ref. 13, p. 47.

137. A. H. Tashjian, Jr. (1979). In ref. 10, p. 529.

138. Y. Yasumura (1968). *Am. Zool.* **8**: 285.

139. F. N. Zeytinoglu, R. F. Gagel, A. H. Tashjian, Jr., R. A. Hammer, and S. E. Leeman (1980). *Proc. Natl. Acad. Sci. USA* **77**: 3741.

140. H. Masui (1981). In ref. 13, p. 109.

141. R. A. Pattillo and G. O. Gey (1968). *Cancer Res.* **28**: 1231.

142. Y. Yasumara, A. H. Tashjian, Jr. and G. H. Sato (1968). *Science* **154**: 1186.

143. S.-I. Shin, Y. Yasumura, and G. H. Sato (1968). *Endocrinology* **82**: 614.

144. G. Augusti-Tocco and G. H. Sato (1969). *Proc. Natl. Acad. Sci. USA* **64**: 311.

145. J. E. Bottenstein (1981). In ref. 13, p. 155.

146. L. A. Greene and A. S. Tischler (1976). *Proc. Natl. Acad. Sci. USA* **73**: 2424.

147. P. Benda, J. Lightbody, G. H. Sato, L. Levine, and W. Sweet (1968). *Science* **161**: 370.

148. N. Sundarraj, M. Schachner, and S. E. Pfeiffer (1975). *Proc. Natl. Acad. Sci. USA* **72**: 1927.

149. J. Lightbody, S. E. Pfeiffer, P. L. Kornblith, and H. Herschman (1970). *J. Neurobiol.* **1**: 411.

150. S. E. Pfeiffer and W. Wechsler (1972). *Proc. Natl. Acad. Sci. USA* **69**: 2885.

151. S. J. Collins, F. W. Ruscetti, R. E. Gallagher, and R. C. Gallo (1978). *Proc. Natl. Acad. Sci. USA* **75**: 2458.

152. Y. Ichikawa (1969). *J. Cell Physiol.* **74**: 223.

153. E. B. Thompson, G. M. Tomkins, and J. F. Curran (1966). *Proc. Natl. Acad. Sci. USA* **56**: 296.

154. U. I. Richardson, A. H. Tashjian, Jr., and L. Levine (1969). *J. Cell Biol.* **40**: 236.

155. U. I. Richardson, P. J. Snodgrass, C. T. Nazum, and A. H. Tashjian, Jr. (1974). *J. Cell Physiol.* **83**: 141.

156. H. C. Pitot, C. Peraino, P. A. Morse, and V. R. Potter (1964). *Natl. Cancer Inst. Monogr.* **13**: 229.

157. J. A. Schneider and M. C. Weiss (1971). *Proc. Natl. Acad. Sci. USA* **68**: 127.

158. C. Szpirer and J. Szpirer (1975). *Differentiation* **4**: 85.

159. D. Cassio and M. C. Weiss (1979). *Somat. Cell Genet.* **5**: 719.

160. G. Moore (1964). *Exp. Cell Res.* **36**: 422.

161. F. Hu and P. F. Lesney (1964). *Cancer Res.* **24**: 1634.

162. T. Kanzaki, T. Kanamaru, S. Nishiyama, H. Eto, H. Kobayashi, and K. Hashimoto (1983). *Dev. Biol.* **99**: 324.

163. D. C. Bennett, L. A. Peachey, H. Durbin, and P. S. Rudland (1978). *Cell* **15**: 283.

164. P. J. Simons, A. A. Tuffey, D. J. McCully, and E. J. Aw (1971). *J. Natl. Cancer Inst.* **46**: 1229.

165. R. Schindler, M. Day, and G. A. Fischer (1959). *Cancer Res.* **19**: 47.

166. N. W. Jessop and R. J. Hay (1980). *In Vitro* **16**: 212.

167. M. D. Walker, T. Edlund, A. M. Boulet, and W. J. Rutter (1983). *Nature* **306**: 557.

168. R. A. Weiss and D. L. Njeuma (1971). In *Growth control in cell cultures* (G. E. W. Wolstenholme and J. Knight, eds.) (Livingstone, London), p. 169.

169. S. J. Gaunt and J. H. Subak-Sharpe (1977). *Exp. Cell Res.* **109**: 341.

170. M. A. Lieberman and L. Glaser (1981). *J. Mem. Biol.* **63**: 1.

171. R. Shields (1976). *Nature* **262**: 348.

172. M. G. P. Stoker (1967). *J. Cell Sci.* **2**: 293.

173. M. Abercrombie (1979). *Nature* **281**: 259.

174. M. L. Hooper (1982). *Biochim. Biophys. Acta* **651**: 85.

175. L. C. M. Reid (1979). In ref. 10, p. 152.

176. A. Lwoff, R. Dulbecco, M. Vogt, and N. Lwoff (1955). *Virology* **1**: 128.

177. L. L. Coriell (1979). In ref. 10, p. 29.

178. R. J. Klebe and M. G. Mancuso (1983). *In Vitro* **19**: 167.

179. H. J. Burki and S. Okada (1968). *Biophys. J.* **8**: 445.

180. A. H. Dantzig, C. W. Slayman, and E. A. Adelberg (1982). *Somat. Cell Genet.* **8**: 509.

181. T. T. Puck and F. T. Kao (1967). *Proc. Natl. Acad. Sci. USA* **58**: 1227.

182. P. S. G. Goldfarb, B. Carritt, M. L. Hooper, and C. Slack (1977). *Exp. Cell Res.* **104**: 357.

183. C. E. Myers (1981). *Pharmacol. Rev.* **33**: 1.

184. R. R. Rueckert and C. C. Mueller (1960). *Cancer Res.* **20**: 1584.

185. M. L. Hooper and R. H. M. Morgan (1979). *Exp. Cell Res.* **119**: 410.

186. C. Basilico (1977). *Adv. Cancer Res.* **24**: 223.

187. H. M. Cann and L. A. Herzenberg (1963). *J. Exp. Med.* **117**: 267.

188. R. R. Porter and K. B. M. Reid (1978). *Nature* **275**: 699.

189. R. Andrews-Wagner and C. H. Sibley (1983). *Somat. Cell Genet.* **9**: 43.

190. D. Pious, P. Hawley, and G. Forrest (1973). *Proc. Natl. Acad. Sci. USA* **70**: 1397.

191. K. G. Sundqvist (1977). In *Dynamic aspects of cell surface organisation*, Cell Surface Reviews, vol. 3 (G. Poste and G. L. Nicolson, eds.) (North Holland, Amsterdam, p. 551).

192. B. W. Papermaster and L. A. Herzenberg (1966). *J. Cell Physiol.* **67**: 407.

193. T. T. Puck, P. Wuthier, C. Jones, and F. T. Kao (1971). *Proc. Natl. Acad. Sci. USA* **68**: 3102.

194. P. Wuthier, C. Jones, and T. T. Puck (1973). *J. Exp. Med.* **138**: 229.

195. C. Jones, P. Wuthier, and T. T. Puck (1975). *Somat. Cell Genet.* **1**: 235.

196. E. S. Vitetta, K. A. Krolick, M. Miyama-Inaba, W. Cushley, and J. W. Uhr (1983). *Science* **219**: 644.

197. G. Moller (1982). *Antibody carriers of drugs and toxins in tumor therapy*, Immunological Reviews, vol. 62 (Munksgaard, Copenhagen).

198. P. E. Thorpe and W. C. J. Ross (1982). *Immunol. Rev.* **62**: 119.

199. V. Raso (1982). *Immunol. Rev.* **62**: 93.

200. C. Milstein and A. C. Cuello (1983). *Nature* **305**: 537.

201. T. Block and M. Bothwell (1983). *Nature* **301**: 342.

202. W. E. Wright (1978). *Exp. Cell Res.* **112**: 395.

203. W. E. Wright (1982). In ref. 6, p. 47.

204. M. L. Hooper and J. H. Subak-Sharpe (1981). *Int. Rev. Cytol.* **69**: 45.

205. H. Wigzell and B. Anderson (1971). *Ann. Rev. Microbiol.* **25**: 291.

206. S. F. Schlossman and L. Hudson (1973). *J. Immunol.* **110**: 313.

207. L. Chess, R. P. MacDermott, and S. F. Schlossman (1974). *J. Immunol.* **113**: 1113.

208. V. Ghetie, G. Mota, and J. Sjoquist (1978). *J. Immunol. Methods* **21**: 133.

209. L. J. Wysocki and V. L. Sato (1978). *Proc. Natl. Acad. Sci. USA* **75**: 2844.

210. N. R. Ling and P. R. Richards (1981). *J. Immunol. Methods* **47**: 265.

211. C. R. Parish (1975). *Transplant Rev.* **25**: 98.

212. L. A. Herzenberg and L. A. Herzenberg (1978). In *Handbook of Experimental Immunology*, 3rd ed. (D. M. Weir, ed.) (Blackwell, Oxford), p. 22.1.

213. B. Arnold, F. L. Battye, and J. F. A. P. Miller (1979). *J. Immonol. Methods* **29**: 353.

214. R. J. Kaufman, J. R. Bertino, and R. T. Schimke (1978). *J. Biol. Chem.* **253**: 5852.

215. H. Goodall and M. Johnson (1982). *Nature* **295**: 524.

216. G. Sablitzky, A. Radbruch, and K. Rajewsky (1982). *Immunol. Rev.* **67**: 59.

217. J. L. Dangl and L. A. Herzenberg (1982). *J. Immunol. Methods* **52**: 1.

218. B. Høltkamp, M. Cramer, H. Lemke, and K. Rajewsky (1981). *Nature* **289**: 66.

219. M. C. Weiss and H. Green (1967). *Proc. Natl. Acad. Sci. USA* **58**: 1104.

220. J. J. Maio and L. de Carli (1962). *Nature* **196**: 600.

221. L. H. Thompson, J. S. Rubin, J. E. Cleaver, G. F. Whitmore, and K. Brookman (1980). *Somat. Cell Genet.* **6**: 391.

222. P. Coffino, R. Baumal, R. Laskov, and M. D. Scharff (1971). *J. Cell Physiol.* **79**: 429.

223. J. L. Preud'homme, B. K. Birshtein, and M. D. Scharff (1975). *Proc. Natl. Acad. Sci. USA* **72**: 1427.

224. J. Sharon, S. L. Morrison, and E. A. Kabat (1979). *Proc. Natl. Acad. Sci. USA* **76**: 1420.

225. N. K . Jerne, C. Henry, A. A. Nordin, H. Fuji, A. M. C. Koros, and I. Lefkovits (1974). *Transplant Rev.* **18**: 130.

226. G. Galfre and C. Milstein (1981). In *Methods in enzymology*, vol. 73 (J. J. Langone and H. van Vunakis, eds.) (Academic Press, New York) p. 3.

227. M. A. Feldman and J. Y. Chou (1983). *In Vitro* **19**: 171.

228. L. L. Cavalli-Sforza and J. Lederberg (1956). *Genetics* **41**: 367.

229. G. Marin (1969). *Exp. Cell Res.* **57**: 29.

230. M. Rosenstraus and L. A. Chasin (1975). *Proc. Natl. Acad. Sci. USA* **72**: 493.

231. W. E. C. Bradley and D. Letovanec (1982). *Somat. Cell Res.* **8**: 51.

232. T. Kuroki (1973). *Exp. Cell Res.* **80**: 55.

233. P. A. Jeggo, L. M. Kemp, and R. Holliday (1982). *Biochimie* **64**: 713.

234. J. Lederberg and E. M. Lederberg (1952). *J. Bacteriol.* **63**: 399.

235. R. A. Goldsby and E. Zipser (1969). *Exp. Cell Res.* **54**: 271.

236. T. D. Stamato and L. K. Hohmann (1975). *Cytogenet. Cell Genet.* **15**: 372.

237. T. D. Stamato and C. A. Waldren (1977). *Somat. Cell Genet.* **3**: 431.

238. J. D. Esko and C. R. H. Raetz (1978). *Proc. Natl. Acad. Sci. USA* **75**: 1190.

239. C. R. H. Raetz, M. M. Wermuth, T. M. McIntyre, J. D. Esko, and D. C. Wing (1982). *Proc. Natl. Acad. Sci. USA* **79**: 3223.

240. M. L. Hooper and R. H. M. Morgan (1979). *Exp. Cell Res.* **123**: 392.

241. R. M. Baker and V. Ling (1978). In *Methods in membrane biology*, vol. 9 (E. D. Korn, ed.) (Plenum, New York), p. 337.

242. S. J. O'Brien, ed. (1982). *Genetic maps*, vol. 2 (National Institutes of Health, Bethesda, Md.).

243. S. L. Naylor, L. L. Busby, and R. J. Klebe (1976). *Somat. Cell Genet.* **2**: 93.

244. B. Carritt, P. S. G. Goldfarb, M. L. Hooper, and C. Slack (1977). *Exp. Cell Res.* **106**: 71.

245. S. L. Naylor, J. K. Townsend, and R. J. Klebe (1979). *Somat. Cell Genet.* **5**: 271.

246. E. C. Cox (1976). *Ann. Rev. Genet.* **10**: 135.

247. P. K. Liu, C. C. Chang, J. E. Trosko, D. K. Dube, G. M. Martin, and L. A. Loeb (1983). *Proc. Natl. Acad. Sci. USA* **80**: 797.

248. D. B. Busch, J. E. Cleaver, and D. A. Glaser (1980). *Somat. Cell Genet.* **6**: 407.

249. R. D. Wood and H. J. Burki (1982). *Mutation Res.* **95**: 505.

250. L. H. Thompson, K. W. Brookman, L. E. Dillehay, C. L. Mooney, and A. V. Carrano (1982). *Somat. Cell Genet.* **8**: 759.

251. V. M. Maher and J. M. McCormick (1980). In *Chemical mutagens* (F. J. de Serres and A. Hollaender, eds.) (Plenum, New York), p. 309.

252. M. Meuth, N. L'Heureux-Huard, and M. Trudel (1979). *Proc. Natl. Acad. Sci. USA* **76**: 6505.

253. M. Meuth, O. Goncalves, and P. Thom (1982). *Somat. Cell Genet.* **8**: 423.

254. E. Drobetsky and M. Meuth (1983). *Mol. Cell Biol.* **3**: 1882.

255. L. A. Wims and S. L. Morrison (1981). *Mutation Res.* **81**: 215.

256. J. W. Drake and R. H. Baltz (1976). *Ann. Rev. Biochem.* **45**: 11.

257. C. F. Arlett (1978). In *Progress in genetic toxicology* (D. Scott, B. A. Bridges, and F. H. Sobels, eds.) (Elsevier-North Holland, Amsterdam), p. 141.

258. J. E. Lever (1976). *J. Cell Physiol.* **89**: 811.

259. T. Gichner and J. Veleminsky (1982). *Mutation Res.* **9**: 129.

260. B. Singer and J. T. Kusmierek (1982). *Ann. Rev. Biochem.* **55**: 655.

261. D. M. Cox and T. T. Puck (1969). *Cytogenetics* **8**: 158.

262. M. W. Berns, J. Aist, J. Edwards, K. Strans, J. Girton, P. McNeil, J. B. Rattner, M. Kitzes, M. Hammer-Wilson, and L. H. Liaw (1981). *Science* **213**: 505.

263. D. C. Wallace (1982). In ref. 6, p. 159.

264. M. Ziegler (1982). In ref. 6, p. 211.

265. W. P. Summers (1973). *Mutation Res.* **20**: 377.

266. D. Clive, R. McCuen, J. F. S. Spector, C. Piper, and K. H. Mavournin (1983). *Mutation Res.* **115**: 225.

267. J. W. Drake (1970). *The molecular basis of mutation* (Holden-Day, San Francisco).

268. C. H. Clarke and M. T. Wade (1975). *Mutation Res.* **28**: 123.

269. B. A. Bridges and R. P. Mottershead (1979). *Mutation Res.* **44**: 305.

270. T. M. Ong and F. J. D. de Serres (1975). *Genetics* **80**: 475.

271. M. J. Hynes (1979). *Genetics* **91**: 381.

272. S. L. Huang, D. N. Rader, and C-Y. Lee (1978). *Chem. Biol. Interactions* **20**: 333.

273. B. A. Bridges and J. Huckle (1970). *Mutation Res.* **10**: 141.

274. E. H. Y. Chu (1971). *Mutation Res.* **11**: 23.

275.   C. F. Arlett and J. Potter (1971). *Mutation Res.* **13**: 59.

276.   J. Thacker and R. Cox (1975). *Nature* **258**: 429.

277.   C. F. Arlett and S. A. Harcourt (1972). *Mutation Res.* **16**: 301.

278.   M. P. Calos and J. H. Miller (1980). *Cell* **20**: 579.

279.   M. I. Marshak, N. B. Varshaver, and N. I. Shapiro (1975). *Mutation Res.* **30**: 383.

280.   H. E. Varmus, N. Quintrell, and S. Ortiz (1981). *Cell* **25**: 23.

281.   G. B. Clements (1975). *Adv. Cancer Res.* **21**: 274.

282.   J. Hochstadt, H. L. Ozer, and C. Schopsis (1981). *Current Topics Microbiol. Immunol.* **94/95**: 243.

283.   T. T. Puck and F-T. Kao (1982). *Ann. Rev. Genet.* **16**: 225.

284.   P. N. Ray and L. Siminovitch (1982). In ref. 5, p. 127.

285.   E. B. Briles (1982). *Int. Rev. Cytol.* **75**: 101.

286.   P. R. Coleman, D. P. Suttle, and G. R. Stark (1977). *J. Biol. Chem.* **252**: 6379.

287.   L. H. Thompson, D. J. Lofgren, and G. M. Adair (1978). *Somat. Cell Genet.* **4**: 423.

288.   C. R. Ashman (1978). *Somat. Cell Genet.* **4**: 299.

289.   M. Fox (1984). *Nature* **307**: 212.

290.   B. Lewin (1980). *Gene expression*, vol. 2, 2nd ed. (Wiley, New York), Chap. 18.

291.   G. M. Wahl, B. Robert de Saint Vincent, and M. L. DeRose (1984). *Nature* **307**: 516.

292.   G. Poste and I. J. Fidler (1980). *Nature* **283**: 139.

293.   F. T. Kao and T. T. Puck (1969). *J. Cell Physiol.* **74**: 245.

294.   F. T. Kao, L. Chasin, and T. T. Puck (1969). *Proc. Natl. Acad. Sci. USA* **64**: 1284.

295.   F. T. Kao and T. T. Puck (1967). *Genetics* **55**: 513.

296.   D. Patterson, F. T. Kao, and T. T. Puck (1975). *Proc. Natl. Acad. Sci. USA* **71**: 2057.

297.   D. Patterson (1975). *Somat. Cell Genet.* **1**: 91.

298.   D. Patterson (1976). *Somat. Cell Genet.* **2**: 41.

299.   D. Patterson (1976). *Somat. Cell Genet.* **2**: 189.

300.   D. C. Oates and D. Patterson (1977). *Somat. Cell Genet.* **3**: 561.

301.   E. H. Y. Chu, N. C. Sun, and C. C. Chang (1972). *Proc. Natl. Acad. Sci. USA* **69**: 3459.

302.   R. K. Feldman and M. W. Taylor (1975). *Biochem.Genet.* **13**: 227.

303.   R. K. Feldman and M. W. Taylor (1975). *Biochem.Genet.* **12**: 393.

304.   E. W. Holmes, G. L.King, A. Layva, and S. C. Singer (1976). *Proc. Natl. Acad. Sci. USA* **73**: 2458.

305.   R. T. Taylor and M. L. Hanna (1977). *Arch. Biochem. Biophys.* **181**: 331.

306.   M. McBurney and G. F. Whitmore (1974). *Cell* **2**: 173.

307.   M. McBurney and G. F. Whitmore (1974). *Cell* **2**: 183.

308.   G. Urlaub and L. A. Chasin (1980). *Proc. Natl. Acad. Sci. USA* **77**: 4216.

309.   D. Ayusawa, H. Koyama, K. Iwata, and T. Seno (1980). *Somat. Cell Genet.* **6**: 261.

310.   D. Patterson and D. V. Carnright (1977). *Somat. Cell Genet.* **3**: 483.

311.   J. N. Davidson, D. V. Carnright, and D. Patterson (1979). *Somat. Cell Genet.* **5**: 176.

312.   T. Kusano, M. Kato, and I. Yamane (1976). *Cell Struct. Func.* **1**: 393.

313. D. Patterson (1980). *Somat. Cell Genet.* **6**: 101.

314. R. S. Krooth, W-L. Hsaio, and B. W. Potvin (1979). *Somat. Cell Genet.* **5**: 551.

315. P. Stamato and D. Patterson (1979). *J. Cell Physiol.* **98**: 459.

316. D. Patterson, D. B. Vannais, and W. Laas (1983). *J. Cell Physiol.* **116**: 257.

317. Y. Saito, S. M. Chou, and D. F. Silbert (1977). *Proc. Natl. Acad. Sci. USA* **74**: 3730.

318. T-Y. Chang, C. Telakowski, W. Vanden Heuvel, A. W. Alberts, and P. R. Vagelos (1977). *Proc. Natl. Acad. Sci. USA* **74**: 832.

319. J. S. Limanek, J. Chu, and T-Y. Chang (1978). *Proc. Natl. Acad. Sci. USA* **75**: 5452.

320. K. Hikada, S-I. Akiyama, and M. Kuwano (1980). *Exp. Cell Res.* **128**: 215.

321. C. Jones and T. T. Puck (1973). *J. Cell Physiol.* **81**: 299.

322. O. Hankinson (1976). *Somat. Cell Genet.* **2**: 497.

323. M. M. Y. Waye and C. P. Stanners (1979). *Somat. Cell Genet.* **5**: 625.

324. G. Ditta, K. Soderberg, F. Landy, and I. E. Scheffler (1976). *Somat. Cell Genet.* **2**: 331.

325. L. DeFrancesco, D. Werntz, and I. E. Scheffler (1975). *J. Cell Physiol.* **85**: 293.

326. T-Y. Chang and P. R. Vagelos (1976). *Proc. Natl. Acad. Sci. USA* **73**: 24.

327. L. H. Thompson, J. L. Harkins, and C. P. Stanners (1973). *Proc. Natl. Acad. Sci. USA* **70**: 3094.

328. L. H. Thompson, C. P. Stanners, and L. Siminovitch (1975). *Somat. Cell Genet.* **1**: 187.

329. R. A. Farber and M. P. Deutscher (1976). *Somat. Cell Genet.* **2**: 509.

330. L. Haars, A. Hampel, and L. H. Thompson (1976). *Biochim. Biophys. Acta* **454**: 493.

331. J. J. Wasmuth and C. T. Caskey (1976). *Cell* **9**: 655.

332. L. H. Thompson, D. J. Lofgren, and G. M. Adair (1977). *Cell* **11**: 157.

333. G. M. Adair, L. H. Thompson, and P. A. Linde (1978). *Somat. Cell Genet.* **4**: 27.

334. I. L. Andrulis, G. S. Chiang, S. M. Arfin, T. A. Miner, and G. W. Hatfield (1978). *J. Biol. Chem.* **253**: 58.

335. C. P. Stanners, T. M. Wightman, and J. L. Harkins (1978). *J. Cell Physiol.* **95**: 125.

336. G. M. Adair, L. H. Thompson, and S. Fond (1978). *Somat. Cell Genet.* **5**: 329.

337. D. Toniolo, H. K. Meiss, and C. Basilico (1973). *Proc. Natl. Acad. Sci. USA* **70**: 1273.

338. D. Toniolo and C. Basilico (1976). *Biochim. Biophys. Acta* **425**: 409.

339. L. H. Thompson, R. Mankovitz, R. M. Baker, J. A. Wright, J. E. Till, L. Siminovitch, and G. F. Whitmore (1971). *J. Cell Physiol.* **78**: 431.

340. M. L. Slater and H. L. Ozer (1976). *Cell* **7**: 289.

341. R. Sheinin (1976). *Cell* **7**: 49.

342. K. K. Jha and H. Ozer (1977). *Genetics (Suppl.)* **86**: 532.

343. D. J. Roufa, S. M. McGill, and J. W. Mollenkamp (1979). *Somat. Cell Genet.* **5**: 97.

344. R. G. Fenwick, Jr. and C. T. Caskey (1975). *Cell* **5**: 115.

345. R. G. Fenwick, Jr., T. H. Sawyer, G. D. Kruh, K. H. Astrin, and C. T. Caskey (1979). *Cell* **12**: 383.

346. C. J. Ingles (1978). *Proc. Natl. Acad. Sci. USA* **75**: 405.

347. M. T. M. Shander, C. Croce, and R. Weinmann (1982). *J. Cell Physiol.* **113**: 324.

348. M. M. Nakano, T. Sekiquicki, and M. Hamada (1978). *Somat. Cell Genet.* **4**: 169.

349. H. C. Renger and C. Basilico (1972). *Proc. Natl. Acad. Sci. USA* **69**: 109.

350. H. C. Renger and C. Basilico (1973). *J. Virol.* **11**: 702.

351. K. D. Noonan, H. C. Renger, C. Basilico, and M. M. Burger (1973). *Proc. Natl. Acad. Sci. USA* **70**: 347.

352. C. Basilico, H. C. Renger, S. J. Burstin, and D. Toniolo (1974). In *Control of proliferation in animal cells* (B. Clarkson and R. Baserga, eds.) (Cold Spring Harbor Press, Cold Spring Harbor, N.Y.), p. 167.

353. N. Yamaguchi and I. B. Weinstein (1975). *Proc. Natl. Acad. Sci. USA* **72**: 214.

354. K. Miyashita and T. Kakunaga (1975). *Cell* **5**: 131.

355. D. Toniolo and C. Basilico (1975). *Cell* **4**: 255.

356. C. H. Schroder and A. W. Hsie (1975). *Exp. Cell Res.* **91**: 170.

357. N. C. Sun, C. C. Chang, and E. H. Y. Chu (1975). *Proc. Natl. Acad. Sci. USA* **72**: 469.

358. V. Ling (1977). *J. Cell Physiol.* **91**: 209.

359. M. C. Willingham, R. A. Carchman, and I. Pastan (1973). *Proc. Natl. Acad. Sci. USA* **70**: 2906.

360. I. E. Scheffler and G. Buttin (1973). *J. Cell Physiol.* **81**: 199.

361. P. M. Naha, A. L. Meyer, and K. Hewitt (1975). *Nature* **258**: 49.

362. R. J. Wang (1976). *Cell* **8**: 257.

363. C. Basilico (1978). *J. Cell Physiol.* **95**: 3677.

364. N. B. Liskay and D. M. Prescott (1978). *Proc. Natl. Acad. Sci. USA* **75**: 2873.

365. H. K. Meiss, A. Talavera, and T. Nishimoto (1978). *Somat. Cell Genet.* **4**: 125.

366. T. Nishimoto and C. Basilico (1978). *Somat. Cell Genet.* **4**: 323.

367. J. Melero (1979). *J. Cell Physiol.* **98**: 17.

368. H. E. Schwartz, G. C. Moser, S. Holmes, and H. K. Meiss (1979). *Somat. Cell Genet.* **5**: 217.

369. P. M. Naha and R. Sorrentino (1980). *Cell Biol. Int. Rep.* **4**: 365.

370. T. Nishimoto, T. Takahashi, and C. Basilico (1980). *Somat. Cell Genet.* **6**: 465.

371. R. T. Schimke and D. A. Haber (1982). In ref. 5, p. 97.

372. M. Meuth and H. Green (1974). *Cell* **3**: 367.

373. W. H. Lewis and J. A. Wright (1979). *Somat. Cell Genet.* **5**: 83.

374. T. Kempe, E. Swyryd, M. Bruist, and G. Stark (1976). *Cell* **9**: 541.

375. G. M. Wahl, R. A. Padgett, and G. R. Stark (1979). *J. Biol. Chem.* **254**: 8679.

376. F. Ardeshir, E. Giulotto, J. Zieg, O. Brison, W. S. L. Liao, and G. R. Stark (1983). *Mol. Cell Biol.* **3**: 2076.

377. J. Zieg, C. E. Clayton, F. Ardeshir, E. Giulotto, E. A. Swyryd, and G. R. Stark (1983). *Mol. Cell Biol.* **3**: 2089.

378. B. B. Levinson, B. Ullman, and D. W. Martin, Jr. (1979). *J. Biol. Chem.* **254**: 4396.

379. D. P. Suttle and G. R. Stark (1979). *J. Biol. Chem.* **254**: 4602.

380. I. L. Andrulis and L. Siminovitch (1981). *Proc. Natl. Acad. Sci. USA* **78**: 5724.

381. J. Choi and I. E. Scheffler (1981). *Somat. Cell Genet.* **7**: 219.

382. M. Debatisse, M. Berry, and G. Buttin (1981). *J. Cell Physiol.* **106**: 1.

383. M. L. Hooper, B. Carritt, P. S. G. Goldfarb, and C. Slack (1977). *Somat. Cell Genet.* **3**: 313.

384. R. H. Wilson (1981). *Heredity* **46**: 285.

385. R. H. Wilson (1982). *Heredity* **49**: 131.

386. L. B. Jacoby (1978). *Somat. Cell Genet.* **4**: 221.

387. L. R. Beach and R. D. Palmiter (1981). *Proc. Natl. Acad. Sci. USA* **78**: 2110.

388. C. T. Caskey and G. D. Kruh (1979). *Cell* **16**: 1.

389. R. G. Fenwick, D. S. Konecki, and C. T. Caskey (1982). In ref. 5, p. 19.

390. L. A. Chasin (1974). *Cell* **2**: 43.

391. G. E. Jones and P. A. Sargent (1974). *Cell* **2**: 37.

392. M. W. Taylor, J. H. Pipkorn, M. K. Tokito, and R. O. Pozzatti, Jr. (1977). *Somat. Cell Genet.* **3**: 195.

393. M. W. McBurney and G. F. Whitmore (1975). *J. Cell Physiol.* **85**: 87.

394. R. S. Gupta and L. Siminovitch (1978). *Somat. Cell Genet.* **4**: 715.

395. T-S. Chan, R. P. Creagan, and M. P. Reardon (1978). *Somat. Cell Genet.* **4**: 1.

396. M. S. Rabin and M. M. Gottesman (1979). *Somat. Cell Genet.* **5**: 571.

397. V. L. Chan and P. Juranka (1981). *Somat. Cell Genet.* **7**: 147.

398. B. Ullman, B. Levinson, M. Hershfield, and D. Martin (1981). *J. Biol. Chem.* **256**: 848.

399. B. Ullman, L. H. Gudas, S. M. Clift, and D. W. Martin, Jr. (1979). *Proc. Natl. Acad. Sci. USA* **76**: 1074.

400. P. Hoffee (1979). *Somat. Cell Genet.* **5**: 319.

401. S. Kit, D. Dubbs, L. H. Piekarski, and T. C. Hsu (1963). *Exp. Cell Res.* **31**: 297.

402. J. W. Littlefield (1965). *Biochim. Biophys. Acta* **95**: 14.

403. C. Slack, R. H. M. Morgan, B. Carritt, P. S. G. Goldfarb, and M. L. Hooper (1976). *Exp. Cell Res.* **98**: 1.

404. D. Clive, K. Johnson, J. Spector, A. Batson, and M. Brown (1979). *Mutation Res.* **59**: 61.

405. T-S. Chan, C. Long, and H. Green (1975). *Somat. Cell Genet.* **1**: 81.

406. G. Buttin, M. Debatisse, and B. Robert de Saint Vincent (1982). In ref. 5, p. 1.

407. L. Medrano and H. Green (1974). *Cell* **1**: 23.

408. M. Greenbert, D. E. Shumm, and T. E. Webb (1979). *Biochem. J.* **164**: 379.

409. C. A. Pasternak, G. A. Fischer, and R. E. Handschumacher (1961). *Cancer Res.* **21**: 110.

410. D. Ayusawa, K. Iwata, and T. Seno (1981). *Somat. Cell Genet.* **7**: 27.

411. E. Arpaia, P. N. Ray, and L. Siminovitch (1983). *Somat. Cell Genet.* **9**: 287.

412. V. Chan, G. F. Whitmore, and L. Siminovitch (1972). *Proc. Natl. Acad. Sci. USA* **69**: 3119.

413. P. E. Lobban and L. Siminovitch (1975). *Cell* **4**: 167.

414. P. E. Lobban and L. Siminovitch (1976). *Cell* **8**: 65.

415. C. J. Ingles, A. Guialis, J. Lam, and L. Siminovitch (1976). *J. Biol. Chem.* **251**: 2729.

416. C. J. Ingles, M. L. Pearson, M. Buchwald, B. C. Beatty, M. M. Crerar, A. Guialis, P. E. Lobban, L. Siminovitch, and D. C. Somers (1976). In *RNA polymerase* (M. Chamberlain and R. Losick, eds.) (Cold Spring Harbor Laboratory, Cold Spring Harbor, N.Y.), p. 835.

417. R. S. Gupta, D. H. Y. Chan, and L. Siminovitch (1978). *J. Cell Physiol.* **97**: 461.

418. R. S. Gupta and L. Siminovitch (1976). *Cell* **9**: 213.

419. R. S. Gupta and L. Siminovitch (1977). *Cell* **10**: 61.

420. R. S. Gupta and L. Siminovitch (1977). *Biochemistry* **16**: 3209.

421. R. S. Gupta and L. Siminovitch (1978). *Somat. Cell Genet.* **4**: 77.

422. R. S. Gupta and L. Siminovitch (1978). *J. Biol. Chem.* **253**: 3978.

423. J. J. Wasmuth, J. M. Hill, and L. S. Vock (1980). *Somat. Cell Genet.* **6**: 495.

424. R. S. Gupta and L. Siminovitch (1978). *Somat. Cell Genet.* **4**: 355.

425. T. J. Moehring and J. M. Moehring (1977). *Cell* **11**: 447.

426. R. S. Gupta and L. Siminovitch (1978). *Somat. Cell Genet.* **4**: 553.

427. J. M. Moehring and T. J. Moehring (1979). *Somat. Cell Genet.* **5**: 453.

428. R. S. Gupta and L. Siminovitch (1980). *Somat. Cell Genet.* **6**: 361.

429. P. Stanley (1983). *Somat. Cell Genet.* **9**: 593.

430. B. Gilfix, J. Rogers, and B. D. Sanwal (1983). *Mol. Cell Biol.* **3**: 2166.

431. T. Sudo and K. Onodera (1979). *J. Cell Physiol.* **101**: 149.

432. C. H. Sibley and G. M. Tomkins (1974). *Cell* **2**: 213.

433. C. H. Sibley and G. M. Tomkins (1974). *Cell* **2**: 221.

434. S. Bourgeois and R. F. Newby (1977). *Cell* **11**: 423.

435. S. Bourgeois and R. F. Newby (1980). In *Hormones and cancer* (S. Iacobelli et al., eds.) (Raven Press, New York), p. 67.

436. R. M. Baker, D. M. Brunette, R. Mankovitz, L. H. Thompson, C. G. Whitmore, L. Siminovitch, and J. E. Till (1974). *Cell* **1**: 9.

437. W. F. Flintoff, S. V. Davidson, and L. Siminovitch (1976). *Somat. Cell Genet.* **2**: 245.

438. W. F. Flintoff, S. M. Spindler, and L. Siminovitch (1976). *In Vitro* **12**: 749.

439. M. Taub and E. Englesberg (1976). *Somat. Cell Genet.* **2**: 441.

440. M. Taub and E. Englesberg (1978). *J. Cell Physiol.* **97**: 477.

441. E. Englesberg, R. Bass, and W. Heiser (1976). *Somat. Cell Genet.* **2**: 411.

442. M. C. Finkelstein, C. W. Slayma, and E. A. Adelberg (1977). *Proc. Natl. Acad. Sci. USA* **74**: 4549.

443. C. E. Campbell, R. A. Gravel, and R. G. Worton (1981). *Somat. Cell Genet.* **7**: 535.

444. J. Mandel and W. F. Flintoff (1978). *J. Cell Physiol.* **97**: 335.

445. C. MacDonald, M. L. Hooper, T. E. J. Buultjens, and B. Carritt (1980). *Exp. Cell Res.* **127**: 277.

446. V. Ling (1975). *Can. J. Genet. Cytol.* **17**: 503.

447. R. L. Juliano and B. Ling (1976). *Biochim. Biophys. Acta* **445**: 152.

448. R. L. Juliano, J. Graves, and V. Ling (1976). *J. Supramol. Struct.* **4**: 521.

449. J. Pouyssegur, A. Franchi, J-C. Salomon, and P. Silvestre (1980). *Proc. Natl. Acad. Sci. USA* **77**: 2698.

450. C. B. Hirschberg, R. M. Baker, M. Perez, L. A. Spencer, and D. Watson (1981). *Mol. Cell Biol.* **1**: 902.

451. R. M. Baker, C. B. Hirschberg, W. A. O'Brien, T. E. Awerbach, and D. Watson (1982). *Methods in enzymology*, vol. 83 (V. Ginzberg, ed.) (Academic Press, New York), p. 444.

452. W. Heiser and E. Englesberg (1979). *Somat. Cell Genet.* **5**: 345.

453. M. Krieger, M. S. Brown, and J. L. Goldstein (1981). *J. Mol. Biol.* **150**: 167.

454. R. S. Gupta (1983). *Cancer Res.* **43**: 1568.

455. V. Ling, J. E. Aubin, A. Chan, and F. Sarangi (1979). *Cell* **18**: 423.

456. F. Cabral, M. E. Sobel, and M. M. Gottesman (1980). *Cell* **20**: 29.

457. R. A. B. Keates, F. Sarangi, and V. Ling (1981). *Proc. Natl. Acad. Sci. USA* **78**: 5638.

458. F. R. Cabral, I. Abraham, and H. M. Gottesman (1981). *Proc. Natl. Acad. Sci. USA* **78**: 4388.

459. F. R. Cabral (1983). *J. Cell Biol.* **97**: 22.

460. F. R. Cabral, L. Wible, S. Brenner, and B. R. Brinkley (1983). *J. Cell Biol.* **97**: 30.

461. A. E. Lagarde and L. Siminovitch (1979). *Somat. Cell Genet.* **5**: 847.

462. T. Lichtor and G. S. Getz (1978). *Proc. Natl. Acad. Sci. USA* **75**: 324.

463. J. M. Eisenstadt and M. C. Kuhns (1982). In ref. 6, p. 111.

464. G. S. Getz and K. L. Kornafel (1982). In ref. 6, p. 139.

465. M. Harris (1978). *Proc. Natl. Acad. Sci. USA* **75**: 5604.

466. R. B. Wallace and K. B. Freeman (1975). *J. Cell Biol.* **65**: 492.

467. A. Wiseman and G. Attardi (1979). *Somat. Cell Genet.* **5**: 241.

468. C-J. Doersen and E. J. Stanbridge (1982). In ref. 6, p. 139.

469. S. J. Mento and L. Siminovitch (1981). *Virology* **111**: 320.

470. J. P. Thirion, D. Banville, and H. Noel (1976). *Genetics* **83**: 137.

471. A. M. Albrecht, J. L. Biedler, and D. J. Hutchison (1972). *Cancer Res.* **32**: 1539.

472. R. C. Jackson and D. Niethammer (1977). *Eur. J. Cancer* **13**: 567.

473. J. H. Goldie, G. Krystal, D. Hartley, G. Gudauskas, and S. Dedhar (1980). *Eur. J. Cancer* **16**: 1539.

474. R. S. Gupta and L. Siminovitch (1980). *J. Cell Physiol.* **102**: 305.

475. J. J. Wasmuth and C. T. Caskey (1976). *Cell* **8**: 71.

476. M. Sinensky (1977). *Biochem. Biophys. Res. Commun.* **78**: 863.

477. R. S. Gupta and M. Hodgson (1981). *Exp. Cell Res.* **132**: 496.

478. T. J. Moehring and J. M. Moehring (1972). *Infect. Immunol.* **6**: 487.

479. M. Merion, P. Schlesinger, R. M. Brooks, J. M. Moehring, T. M. Moehring, and W. S. Sly (1983). *Proc. Natl. Acad. Sci. USA* **80**: 5315.

480. C. D. Green and D. W. Martin, Jr. (1973). *Proc. Natl. Acad. Sci. USA* **70**: 3698.

481. J. M. Pouyssegur and I. Pastan (1976). *Proc. Natl. Acad. Sci. USA* **73**: 544.

482. J. M. Pouyssegur, M. Willingham, and I. Pastan (1977). *Proc. Natl. Acad. Sci. USA* **74**: 243.

483. R. J. Klebe, P. G. Rosenberger, S. L. Naylor, R. L. Burns, R. Novak, and H. Kleinman (1977). *Exp. Cell Res.* **104**: 119.

484. E. D. Wright, P. S. G. Goldfarb, and J. H. Subak Sharpe (1976). *Exp. Cell Res.* **103**: 63.

485. C. Slack, R. H. M. Morgan, and M. L. Hooper (1978). *Exp. Cell Res.* **117**: 195.

486. K. Willecke, D. Muller, P. M. Druge, U. Frixen, R. Schafer, R. Dermietzel, and D. Hulser (1983). *Exp. Cell Res.* **144**: 95.

487. T. A. Smith and M. L. Hooper (1982). *Biol. Cell* **45**: 88.

488. C. Jones and E. E. Moore (1976). *Somat. Cell Genet.* **2**: 235.

489. S. Salzberg, D. H . Wreschner, F. Oberman, A. Panet, and M. Bakhanashvili (1983). *Mol. Cell Biol.* **3**: 1759.

490. J. D. Esko and C. R. H. Raetz (1980). *Proc. Natl. Acad. Sci. USA* **77**: 5192.

491. M. A. Polokoff, D. C. Wing, and C. R. H. Raetz (1981). *J. Biol. Chem.* **256**: 7687.

492. A. R. Robbins (1979). *Proc. Natl. Acad. Sci. USA* **76**: 1911.

493. J. R. Gum, Jr. and C. R. H. Raetz (1983). *Proc. Natl. Acad. Sci. USA* **80**: 3918.

494. B. Sanwal (1979). *Trends Biochem. Sci.* **4**: 155.

495. W. F. Loomis, Jr., J. P. Wahrmann, and D. Luzzati (1973). *Proc. Natl. Acad. Sci. USA* **70**: 425.

496. K. Adetugbo, C. Milstein, and D. S. Secher (1977). *Nature* **265**: 299.

497. P. Coffino and M. D. Scharff (1971). *Proc. Natl. Acad. Sci. USA* **68**: 219.

498. R. G. H. Cotton, D. S. Secher, and C. Milstein (1973). *Eur. J. Immunol.* **3**: 135.

499. P. Coffino, H. R. Bourne, P. A. Insel, K. L. Melmon, G. Johnson, and J. Vigne (1978). *In Vitro* **14**: 140.

500. T. Haga, E. M. Ross, H. J. Anderson, and H. G. Gilman (1977). *Proc. Natl. Acad. Sci. USA* **74**: 2016.

501. H. Bourne, P. Coffino, and G. M. Tomkins (1975). *Science* **187**: 750.

502. V. Daniel, H. R. Bourne, and G. M. Tomkins (1973). *Nature New Biol.* **244**: 167.

503. P. A. Insel, H. R. Bourne, P. Coffino, and G. M. Tomkins (1975). *Science* **190**: 896.

504. P. Coffino, H. R. Bourne, and G. M. Tomkins (1975). *J. Cell Physiol.* **85**: 603.

505. O. R. Burrone, F. Calabi, R. F. Kefford, and C. Milstein (1983). *EMBO J.* **2**: 1591.

506. J. E. Buss, J. E. Kindlow, C. S. Lazar, and G. N. Gill (1982). *Proc. Natl. Acad. Sci. USA* **79**: 2574.

507. J. Deschatrette and M. C. Weiss (1974). *Biochimie* **56**: 1603.

508. F. Forquignon and B. Ephrussi (1979). *Somat. Cell Genet.* **5**: 409.

509. J. Pawelek, M. Sansone, J. Morowitz, G. Moellman, and E. Godawska (1974). *Proc. Natl. Acad. Sci. USA* **71**: 1073.

510. J. Pawelek, M. Sansone, N. Koch, G. Christie, R. Halaban, J. Hendee, A. B. Lerner, and J. M. Varga (1975). *Proc. Natl. Acad. Sci. USA* **72**: 951.

511. J. W. Littlefield (1964). *Science* **145**: 709.

512. M. Harris (1971). *J. Cell Physiol.* **78**: 177.

513. L. Mezger-Freed (1972). *Nature New Biol.* **235**: 245.

514. S. E. Luria and M. Delbruck (1943). *Genetics* **28**: 491.

515. M. Cappechi, R. A. von der Haar, N. E. Cappechi, and M. M. Sveda (1977). *Cell* **12**: 371.

516. C. Coleclough, R. P. Perry, K. Karjalainen, and P. T. Weigert (1981). *Nature* **290**: 371.

517. K. Adetugbo, C. Milstein, and D. S. Secher (1977). *Nature* **265**: 299.

518. T. Akera (1977). *Science* **198**: 569.

519. J. E. Lever and J. E. Seegmiller (1976). *J. Cell Physiol.* **88**: 343.

520. W. N. Choy and J. W. Littlefield (1980). *Proc. Natl. Acad. Sci. USA* **77**: 1101.

521. G. Dover (1982). *Nature* **299**: 111.

522. R. DeMars (1974). *Mutation Res.* **24**: 335.

523. L. Chasin (1973). *J. Cell Physiol.* **82**: 299.

524. L. Chasin and G. Urlaub (1975). *Science* **187**: 1091.

525. C. E. Campbell and R. G. Worton (1979). *Somat. Cell Genet.* **5**: 51.

526. L. Siminovitch (1979). In *Mammalian cell mutagenesis: The maturation of test systems* (A. H. Hsu, J. P. O'Neill, and V. K. McElheny, eds.) (Cold Spring Harbor Laboratory, Cold Spring Harbor, N.Y.), p. 15.

527. L. Siminovitch (1976). *Cell* **7**: 1.

528. R. S. Gupta, D. H. Y. Chan, and L. Siminovitch (1978). *Cell* **14**: 1007.

529. R. G. Worton, C. Duff, and C. E. Campbell (1980). *Somat. Cell Genet.* **6**: 199.

530. R. S. Gupta (1980). *Somat. Cell Genet.* **6**: 115.

531. M. J. Siciliano, J. Siciliano, and R. M. Humphrey (1978). *Proc. Natl. Acad. Sci. USA* **75**: 1919.

532. E. M. Eves and R. A. Farber (1981). *Proc. Natl. Acad. Sci. USA* **78**: 1768.

533. D. J. Roufa, B. N. Sadow, and C. T. Caskey (1973). *Genetics* **75**: 515.

534. A. E. Simon, M. W. Taylor, and W. E. C. Bradley (1983). *Mol. Cell Biol.* **3**: 1703.

535. G. M. Adair, R. L. Stalling, R. S. Nairn, and M. J. Siciliano (1983). *Proc. Natl. Acad. Sci. USA* **80**: 5961.

536. G. Barski, S. Sorieul, and F. Cornefert (1961). *J.Natl. Cancer Inst.* **26**: 1269.

537. H. Harris and J. F. Watkins (1965). *Nature* **205**: 640.

538. G. Pontecorvo (1976). *Somat. Cell Genet.* **1**: 397.

539. W. E. Mercer and R. Baserga (1982). In ref. 6, p. 23.

540. R. J. Klebe and M. G. Mancuso (1981). *Somat. Cell Genet.* **7**: 473.

541. T. H. Norwood and C. J. Zeigler (1982). In ref. 6, p. 35.

542. R. J. Klebe and M. G. Mancuso (1982). *Somat. Cell Genet.* **8**: 723.

543. G. Buttin, G. LeGuern, L. Phalente, E. C. C. Linn, L. Medrang, and P. A. Cazenave (1978). *Current Topics Microbiol. Immunol.* **81**: 27.

544. U. Zimmerman (1982). *Biochim. Biophys. Acta* **694**: 227.

545. J. Teissié, V. Knudson, T. Y. Tsong, and D. Lane (1982). *Science* **216**: 537.

546. B. Claude and T. Justin (1983). *Biochem. Biophys. Res. Commun.* **114**: 663.

547. H. Galjaard (1982). In *Methods in cell biology*, vol. 26 (S. A. Latt and G. J. Darlington, eds.) (Academic Press, New York), p. 241.

548. D. Bootsma and H. Galjaard (1979). In *Methods for the study of inborn errors of metabolism* (F. A. Hommes, ed.) (Elsevier/North-Holland, Amsterdam), p. 241.

549. P. N. Rao, R. T. Johnson, and K. Sperling, eds. (1982). *Premature chromosome condensation: Application in basic, clinical and maturation research* (Academic Press, New York).

550. W. E. Wright (1981). *J. Cell Biol.* **91**: 11.

551. R. M. Liskay and D. Patterson (1978). In *Methods in cell biology*, vol. 20 (D. M. Prescott, ed.). (Academic Press, New York), p. 335.

552. K. K. Jha and H. L. Ozer (1976). *Somat. Cell Genet.* **2**: 215.

553. M. C. Weiss (1982). In ref. 5, p. 169.

554. M. Mével-Ninio and M. C. Weiss (1981). *J. Cell Biol.* **90**: 339.

555. R. Bertolotti (1977). *Somat. Cell Genet.* **3**: 365.

556. J. Deschatrette, E. E. Moore, M. Dubois, D. Cassio, and M. C. Weiss (1979). *Somat. Cell Genet.* **5**: 697.

557. M. Allan and P. Harrison (1980). *Cell* **19**: 437.

558. D. E. Axelrod, T. V. Gopalakrishnan, M. Willing, and W. F. Anderson (1978). *Somat. Cell Genet.* **4**: 157.

559. C. Fougère and M. C. Weiss (1978). *Cell* **15**: 843.

560. J. Deschatrette and M. C. Weiss (1975). *Somat. Cell Genet.* **1**: 279.

561. G. Kohler and C. Milstein (1975). *Nature* **256**: 495.

562. C. L. Reading (1982). *J. Immunol. Methods* **53**: 261.

563. B. B. Beezley and N. H. Ruddle (1982). *J. Immunol. Methods* **52**: 269.

564. H. v. Boehmer, W. Haas, G. Kohler, F. Melchers, and J. Zeuthen, eds. *Current Topics Microbiol. Immunol.* **100**, whole issue.

565. A. Altman and D. H. Katz (1982). *Adv. Immunol.* **33**: 73.

566. M. B. Mokyr and S. Dray (1982). In *Tumor markers* (H. Busch and L. C. Yeoman, eds.), Methods in Cancer Research, vol. 19, p. 385 (Academic Press, New York).

567. E. Kedar and D. W. Weiss (1984). In *Immunogenicity*, 2nd ed. (F. Borek, ed.) (Elsevier/North Holland, Amsterdam), in press.

568. H. L. Ozer and K. K. Jha (1977). *Adv. Cancer Res.* **25**: 53.

569. K. Willecke and R. Schafer (1982). In ref. 5, p. 183.

570. J. Jonasson and H. Harris (1979). *J. Cell Sci.* **24**: 255.

571. A. N. Howell and R. Sager (1978). *Proc. Natl. Acad. Sci. USA* **75**: 2358.

572. C. L. Bunn (1982). In ref. 6, p. 189.

573. D. D. Pravtcheva and F. H. Ruddle (1983). *Exp. Cell Res.* **146**: 401.

574. D. D. Pravtcheva and F. H. Ruddle (1983). *Exp. Cell Res.* **148**: 265.

575. P. L. Pearson, T. H. Roderick, M. T. Davisson, P. A. Lalley, and S. J. O'Brien (1982). *Cytogenet. Cell Genet.* **32**: 208.

576. S. D. Handmaker (1971). *Nature* **233**: 416.

577. T. B. Shows, A. Y. Sakaguchi, and S. L. Naylor (1982). *Adv. Human Genet.* **12**: 341.

578. S. J. O'Brien, J. M. Simonson, and M. Eichelberger (1982). In ref. 6, p. 513.

579. R. Bravo, S. J. Fey, H. Macdonald-Bravo, R. Schafer, K. Willecke, and J. E. Celis (1982). In ref. 5, p. 43.

580. F. H. Ruddle (1981). *Nature* **294**: 115.

581. A. S. Henderson (1982). *Int. Rev. Cytol.* **76**: 1.

582. S. J. O'Brien and W. G. Nash (1982). *Science* **216**: 257.

583. S. Ohno (1973). *Nature* **244**: 259.

584. P. D'Eustachio and F. H. Ruddle (1983). *Science* **220**: 919.

585. G. E. Veomett (1982). In ref. 6, p. 67.

586. J. F. Jongkind and A. Verkerk (1982). In ref. 6, p. 81.

587. M. A. Clark and J. W. Shay (1982). In ref. 6, p. 269.

588. L. G. Weide, M. A. Clark, and J. W. Shay (1982). In ref. 6, p. 281.

589. R. E. K. Fournier (1982). In ref. 6, p. 309.

590. A. H. Crenshaw, Jr. and L. R. Murrell (1982). In ref. 6, p. 291.

591. R. T. Johnson and A. M. Mullinger (1982). In ref. 6, p. 329.

592. P. N. Goodfellow, G. Banting, G. Trowsdale, S. Chambers, and E. Solomon (1982). *Proc. Natl. Acad. Sci. USA* **79**: 1190.

593. M. J. Hightower and J. J. Lucas (1982). In ref. 6, p. 255.

594. S. J. Goss and H. Harris (1975). *Nature* **255**: 680.

595. S. J. Goss (1979). *Cytogenet. Cell Genet.* **25**: 1.

596. O. W. McBride and J. L. Peterson (1980). *Ann. Rev. Genet.* **14**: 321.

597. O. W. McBride (1982). In ref. 6, p. 375.

598. L. A. Klobutcher and F. H. Ruddle (1979). *Nature* **280**: 657.

599. M. H. McCutchan and J. S. Pagano (1968). *J. Natl. Cancer Inst.* **41**: 351.

600. F. L. Graham and A. J. Van der Eb (1973). *Virology* **52**: 456.

601. N. D. Stow and N. Wilkie (1976). *J. Gen. Virol.* **33**: 447.

602. E. Frost and J. Williams (1978). *Virology* **91**: 39.

603. G. Scangos and F. H. Ruddle (1981). *Gene* **14**: 1.

604. A. P. Bollon and S. J. Silverstein (1982). In ref. 6, p. 415.

605. F. Colbere-Garapin, A. Garapin, and P. Kourilsky (1981). *Current Topics Microbiol. Immunol.* **96**: 145.

606. A. Graessmann and M. Graessmann (1982). In ref. 6, p. 463.

607. F. Yamamoto, M. Furusawa, I. Furusawa, and M. Obinata (1982). *Exp. Cell Res.* **142**: 79.

608. C. W. Lo (1983). *Mol. Cell Biol.* **3**: 1803.

609. R. M. Straubinger and D. Papahadjopoulos (1982). In ref. 6, p. 399.

610. W. Schaffner (1980). *Proc. Natl. Acad. Sci. USA* **77**: 2163.

611. M. Rassoulzadegan, B. Binetruy, and F. Cuzin (1982). *Nature* **295**: 257.

612. B. Robert de Saint Vincent, S. Delbruck, W. Eckhart, J. Meinkoth, L. Vitto, and G. Wahl (1981). *Cell* **27**: 267.

613. H. G. Coon and C. Ho (1978). In *Genetic interactions and gene transfer*, Brookhaven Symposium, vol. 29 (C. W. Anderson, ed.) (Brookhaven National Laboratory, New York), p. 166.

614. M. A. Clark, T. L. Reudelhuber, and J. W. Shay (1982). In ref. 6, p. 203.

615. M. C. Rechsteiner (1982). In ref. 6, p. 385.

616. C. Y. Okada and M. C. Rechsteiner (1982). *Cell* **29**: 33.

617. D. J. Jolly, A. C. Esty, H. U. Bernard, and T. Friedman (1982). *Proc. Natl. Acad. Sci. USA* **79**: 5038.

618. K. M. Huttner, G. Scangos, and F. H. Ruddle (1979). *Proc. Natl. Acad. Sci. USA* **76**: 5820.

619. T. Maniatis, E. F. Fritsch, and J. Sambrook (1982). *Molecular cloning: A laboratory manual* (Cold Spring Harbor Laboratory, Cold Spring Harbor, N.Y.).

620. M. Perucho, D. Hanahan, L. Lipsich, and M. Wigler (1980). *Nature* **285**: 207.

621. I. Lowy, A. Pellicer, J. F. Jackson, G-K. Sim, S. Silverstein, and R. Axel (1980). *Cell* **22**: 817.

622. J. H. Nunberg, R. J. Kaufman, A. C. Y. Chang, S. N. Cohen, and R. T. Schimke (1980). *Cell* **19**: 355.

623. P. W. J. Rigby (1982). In *Genetic engineering*, vol. 3 (R. Williamson, ed.) (Academic Press, New York), p. 83.

624. P. W. J. Rigby (1983). *J. Gen. Virol.* **64**: 255.

625. R. Mann, R. C. Mulligan, and D. Baltimore (1983). *Cell* **33**: 153.

626. S. Watanabe and H. M. Temin (1983). *Mol. Cell Biol.* **3**: 2241.

627. H. Varmus and R. Swanstrom (1982). In *RNA tumor viruses* (R. Weiss, N. Teich, H. Varmus, and J. Coffin, eds.) (Cold Spring Harbor Laboratory, Cold Spring Harbor, N.Y.), p. 369.

628. Y. Gluzman (1981). *Cell* **23**: 175.

629. A. Razzaque, H. Mizusawa, and M. M. Seidman (1983). *Proc. Natl. Acad. Sci. USA* **80**: 3010.

630. M. P. Calos, J. S. Lebkowski, and M. R. Botchan (1983). *Proc. Natl. Acad. Sci. USA* **80**: 3015.

631. J. E. Darnell, Jr. (1982). *Nature* **297**: 365.

632. C. Montell, E. F. Fisher, M. H. Caruthers, and A. J. Berk (1983). *Nature* **305**: 600.

633. S. A. Mitsalis, J. F. Young, P. Patesse, and R. V. Guntaka (1981). *Gene* **16**: 217.

634. R. M. Hudziak, F. A. Laski, U. L. RajBhandary, P. A. Sharp, and M. R. Capecchi (1982). *Cell* **31**: 137.

635. R. L. Brinster, H. Y. Chen, M. Trumbauer, A. W. Senear, R. Warren, and R. D. Palmiter (1981). *Cell* **27**: 223.

636. K. E. Mayo, R. Warren, and R. D. Palmiter (1982). *Cell* **29**: 99.

637. K. E. Mercola, H. D. Stang, J. Browne, W. Salser, and M. J. Cline (1980). *Science* **208**: 1033.

638. D. Schumperli, B. H. Howard, and M. R. Rosenberg (1982). *Proc. Natl. Acad. Sci. USA* **79**: 257.

639. R. C. Mulligan and P. Berg (1981). *Proc. Natl. Acad. Sci. USA* **78**: 2072.

640. R. C. Mulligan and P. Berg (1980). *Science* **209**: 1422.

641. J. F. Nicolas and P. Berg (1983). In ref. 711, p. 469.

642. M. Karin, G. Cathala, and Chi Nguyen-Huu (1983). *Proc. Natl. Acad. Sci. USA* **80**: 4040.

643. S. Subramani and P. J. Southern (1983). *Anal. Biochem.* **135**: 1.

644. C. M. Gorman, L. F. Moffat, and B. H. Howard (1982). *Mol. Cell Biol.* **2**: 1044.

645. M. Yaniv (1982). *Nature* **297**: 18.

646. H. Weiher, M. Konig, and P. Gruss (1983). *Science* **219**: 626.

647. F. Lee, R. Mulligan, P. Berg, and G. Ringold (1981). *Nature* **294**: 228.

648. H. Okayama and P. Berg (1983). *Mol. Cell Biol.* **3**: 280.

649. H. Okayama and P. Berg (1982). *Mol. Cell Biol.* **2**: 161.

650. D. Di Maio, R. Treisman, and T. Maniatis (1982). *Proc. Natl. Acad. Sci. USA* **79**: 4030.

651. N. Sarver, J. C. Byrne, and P. M. Howley (1982). *Proc. Natl. Acad. Sci. USA* **79**: 7147.

652. P. J. Kushner, B. B. Levinson, and H. M. Goodman (1982). *J. Mol. Appl. Genet.* **1**: 527.

653. M-F. Law, J. C. Byrne, and P. M. Howley (1983). *Mol. Cell Biol.* **3**: 2110.

654. A. Joyner, G. Keller, R. A. Phillips, and A. Bernstein (1983). *Nature* **305**: 556.

655. R. J. Kaufman and P. A. Sharp (1982). *J. Mol. Biol.* **159**: 601.

656. J. D. Milbrandt, J. C. Azizkhan, K. S. Griesen, and J. L. Hamlin (1983). *Mol. Cell Biol.* **3**: 1266.

657. J. D. Milbrandt, J. C. Azizkhan, and J. L. Hamlin (1983). *Mol. Cell Biol.* **3**: 1274.

658. F. G. Grosveld, T. Lund, E. J. Murray, A. L. Mellor, H. H. M. Dahl, and R. A. Flavell (1982). *Nucl. Acids Res.* **10**: 6715.

659. K. Shimotohno and H. Temin (1981). *Cell* **26**: 67.

660. P. J. Southern and P. Berg (1982). *J. Mol. Appl. Genet* **1**: 327.

661. S. Subramani, R. Mulligan, and P. Berg (1981). *Mol. Cell Biol.* **1**: 854.

662. B. H. Howard (1983). *Trends Biochem. Sci.* **8**: 209.

663. A. B. Chapman, M. A. Costello, F. Lee, and G. M. Ringold (1983). *Mol. Cell Biol.* **3**: 1421.

664. F. Colbere-Garapin, F. Horodniceanu, P. Kourilsky, and A. C. Garapin (1981). *J. Mol. Biol.* **150**: 1.

665. A. Linnenbach and C. M. Croce (1982). In ref. 6, p. 429.

666. P. Mellon, V. Parker, Y. Gluzman, and T. Maniatis (1981). *Cell* **27**: 279.

667. K. O'Hare, C. Benoist, and R. Breathnach (1981). *Proc. Natl. Acad. Sci. USA* **78**: 1527.

668. W. Chia, M. R. D. Scott, and P. W. J. Rigby (1982). *Nucl. Acids Res.* **10**: 2503.

669. P. D. Matthias, H. U. Bernard, A. Scott, G. Brady, T. Hashimoto-Gotoh, and G. Schutz (1983). *EMBO J* **2**: 1487.

670. C. J. Tabin, J. W. Hoffmann, S. P. Goff, and R. A. Weinberg (1982). *Mol. Cell Biol.* **2**: 426.

671. C. M. Wei, M. Gibson, P. G. Spear, and E. M. Scolnick (1981). *J. Virol.* **39**: 935.

672. A. Joyner and A. Bernstein (1983). *Mol. Cell Biol.* **3**: 2180.

673. A. D. Miller, D. J. Jolly, T. Friedmann, and I. M. Verma (1983). *Proc. Natl. Acad. Sci. USA* **80**: 4709.

674. R. S. Roginsky, A. I. Skoultchi, P. Henthorn, O. Smithies, N. Hsiung, and R. Kucherlapati (1983). *Cell* **35**: 149.

675. D. T. Kurtz (1981). *Nature* **291**: 629.

676. D. M. Robins, J. Paek, P. Seeburg, and R. Axel (1982). *Cell* **29**: 623.

677. J. Banerji, S. Rusconi, and W. Schaffner (1981). *Cell* **27**: 299.

678. M. V. Chao, P. Mellon, P. Charney, T. Maniatis, and R. Axel (1983). *Cell* **32**: 483.

679. C. Queen and D. Baltimore (1983). *Cell* **33**: 741.

680. J. Stafford and C. Queen (1983). *Nature* **306**: 77.

681. S. D. Gillies, S. L. Morrison, V. T. Oi, and S. Tonegawa (1983). *Cell* **33**: 717.

682. M. S. Neuberger (1983). *EMBO J.* **2**: 1373.

683. J. Banerji, L. Olson, and W. Schaffner (1983). *Cell* **33**: 729.

684. M. A. Boss (1983). *Nature* **303**: 281.

685. B. L. M. Hogan (1983). *Nature* **306**: 313.

686. D. Shortle, D. Di Maio, and D. Nathans (1981). *Ann. Rev. Genet.* **15**: 265.

687. J. E. Celis and J. D. Smith, eds. (1979). *Nonsense mutations and tRNA suppression* (Academic Press, London).

688. F. A. Laski, R. Belagaje, U. L. RajBhandary, and P. A. Sharp (1982). *Proc. Natl. Acad. Sci. USA* **79**: 5813.

689. G. F. Temple, A. M. Dozy, K. L. Roy, and Y. W. Kan (1982). *Nature* **296**: 537.

690. K. Struhl (1983). *Nature* **305**: 391.

691. H. Land, L. F. Parada, and R. A. Weinberg (1983). *Science* **222**: 771.

692. C. Shih, B-Z. Shilo, M. P. Goldfarb, A. Dannenberg, and R. A. Weinberg (1979). *Proc. Natl. Acad. Sci. USA* **76**: 5714.

693. S. Pulciani, E. Santos, A. V. Lauver, L. K. Long, S. A. Aaronson, and M. Barbacid (1982). *Nature* **300**: 539.

694. C. J. Tabin, S. M. Bradley, C. I. Bargmann, R. A. Weinberg, A. G. Papageorge, E. M. Scolnick, R. Dhar, D. R. Lowy, and E. H. Chang (1982). *Nature* **300**: 143.

695. E. P. Reddy, R. K. Reynolds, E. Santos, and M. Barbacid (1982). *Nature* **300**: 149.

696. Y. Yuasa, S. K. Srivastava, C. Y. Dunn, J. S. Rhim, E. P. Reddy, and S. A. Aaronson (1983). *Nature* **303**: 775.

697. H. M. Temin (1983). *Nature* **302**: 656.

698. J. D. Rowley (1983). *Nature* **301**: 290.

699. J. Logan and J. Cairns (1983). *Nature* **300**: 104.

700. J. Cairns and J. Logan (1983). *Nature* **304**: 582.

701. H. Land, L. F. Parada, and R. A. Weinberg (1983). *Nature* **304**: 596.

702. H. E. Ruley (1983). *Nature* **304**: 602.

703. R. F. Newbold and R. W. Overell (1983). *Nature* **304**: 648.

704. P. Newmark (1983). *Nature* **305**: 470.

705. M. D. Waterfield, G. T. Scrace, N. Whittle, P. Stroobant, A. Johnsson, A. Wasteson, B. Westermark, C-H. Heldin, J. S. Huang, and T. F. Deuel (1983). *Nature* **304**: 35.

706. R. F. Doolittle, M. W. Hunkapiller, L. E. Hood, S. G. Devare, K. C. Robbins, S. A. Aaronson, and H. N. Antoniades (1983). *Science* **221**: 275.

707. R. A. Weiss (1983). *Nature* **304**: 12.

708. D. Solter and I. Damjanov (1979). In *Methods in cancer research*, vol. 18 (W. H. Fischman and H. Busch, eds.) (Academic Press, New York), p. 277.

709. G. R. Martin (1980). *Science* **209**: 768.

710. B. Mintz and R. A. Fleischman (1981). *Adv. Cancer Res.* **34**: 211.

711. L. M. Silver, G. R. Martin, and S. Strickland, eds. (1983). *Teratocarcinoma stem cells*, Cold Spring Harbor Conferences on Cell Proliferation, vol. 10 (Cold Spring Harbor Laboratory, Cold Spring Harbor, N.Y.)

712. R. L. Gardner, ed. (1983). *Embryonic and germ cell tumours in man and animals*, Cancer Surveys, vol. 2, no. 1 (Oxford University Press, Oxford, U.K.).

713. K. M. Grigor (1981). *Int. J. Androl. Suppl.* **4**: 35.

714. M. Vandeputte, H. Sobis, A. Billiau, B. van de Maele, and R. Leyten (1973). *Int. J. Cancer* **11**: 536.

715. H. Sobis and M. Vandeputte (1982). *Dev. Biol.* **92**: 553.

716. G. R. Martin (1983). In ref. 711, p. 690.

717. R. Oshima (1978). *Differentiation* **11**: 149.

718. T. A. Smith and M. L. Hooper (1983). *Exp. Cell Res.* **145**: 457.

719. S. Strickland and V. Mahdavi (1978). *Cell* **15**: 393.

720. M. W. McBurney, E. M. V. Jones-Villeneuve, M. K. S. Edwards, and M. Rudnicki (1983). In ref. 711, p. 121.

721. M. K. S. Edwards and M. W. McBurney (1983). *Dev. Biol.* **98**: 187.

722. S. Strickland, K. K. Smith, and K. R. Marotti (1980). *Cell* **21**: 347.

723. B. L. M. Hogan and A. Taylor (1981). *Nature* **291**: 235.

724. Y. Ogiso, A. Kume, Y. Nishimune, and A. Matsushito (1982). *Exp. Cell Res.* **137**: 365.

725. M. A. Eglitis and M. I. Sherman (1983). *Exp. Cell Res.* **146**: 289.

726. M. I. Sherman, M. L. Paternoster, and M. Taketo (1983). *Cancer Res.* **43**: 4283.

727. H. Jacob, P. Dubois, H. Eisen, and F. Jacob (1979). *C.R. Acad. Sci. Ser. D (Paris)* **286**: 109.

728. W. C. Speers, C. R. Birdwell, and F. J. Dixon (1979). *Am. J. Path.* **97**: 563.

729. M. W. McBurney, E. M. V. Jones-Villeneuve, M. K. S. Edwards, and P. J. Anderson (1982). *Nature* **299**: 165.

730. C. Cremisi (1983). *J. Cell Physiol.* **116**: 181.

731. W. C. Speers and J. M. Lehman (1976). *J. Cell Physiol.* **88**: 297.

732. J. Zeuthen (1981). *Int. J. Androl. Suppl.* **4**: 61.

733. R. H. M. Morgan, J. A. Henry, and M. L. Hooper (1983). *Exp. Cell Res.* **148**: 461.

734. C. L. Stewart (1982). *J. Embryol. Exp. Morph.* **67**: 167.

735. V. Papaioannou and J. Rossant (1983). In ref. 711, p. 734.

736. T. A. Stewart and B. Mintz (1982). *J. Exp. Zool.* **224**: 465.

737. J. Rossant and M. W. McBurney (1982). *J. Embryol. Exp. Morph.* **70**: 99.

738. R. S. Wells (1982). *Cancer Res.* **42**: 2736.

739. L. D. Siracusa, V. M. Chapman, K. L. Bennett, N. D. Hastie, D. F. Pietras, and J. Rossant (1983). *J. Embryol. Exp. Morph.* **73**: 163.

740. J. Rossant, M. Vijh, L. D. Siracusa, and V. M. Chapman (1983). *J. Embryol. Exp. Morph.* **73**: 179.

741. J. Rossant and V. M. Chapman (1983). *J. Embryol. Exp. Morph.* **73**: 193.

742. M. J. Evans and M. H. Kaufman (1981). *Nature* **292**: 154.

743. G. R. Martin (1981). *Proc. Natl. Acad. Sci. USA* **78**: 7634.

744. M. J. Evans, E. R. Robertson, A. H. Handyside, and M. H. Kaufman (1982). *Biol. Cell* **45**: 406.

745. H. R. Axelrod and E. Lader (1983). In ref. 711, p. 665.

746. T. Magnuson, C. J. Epstein, L. M. Silver, and G. R. Martin (1982). *Nature* **298**: 750.

747. M. H. Kaufman, E. J. Robertson, A. H. Handyside, and M. J. Evans (1983). *J. Embryol. Exp. Morph.* **73**: 249.

748. E. J. Robertson, M. J. Evans, and M. H. Kaufman (1983). *J. Embryol. Exp. Morph.* **74**: 297.

749. J. K. Heath (1983). In ref. 712, p. 142.

750. A. Rizzino and G. Sato (1978). *Proc. Natl. Acad. Sci. USA* **75**: 1844.

751. A. Rizzino and C. Crowley (1980). *Proc. Natl. Acad. Sci. USA* **77**: 457.

752. A. Rizzino, V. Terranova, D. Rohrbach, C. Crowley, and H. Rizzino (1980). *J. Supramol. Struct.* **13**: 243.

753. M. Y. Darmon (1982). *In Vitro* **18**: 977.

754. A. Rizzino (1983). *Dev. Biol.* **95**: 126.

755. M. Darmon, J. Bottenstein, and G. Sato (1981). *Dev. Biol.* **85**: 463.

756. M. Darmon, W. B. Stallcup, and Q. J. Pittman (1982). *Exp. Cell Res.* **138**: 73.

757. M. Darmon, M. H. Buc-Caron, D. Paulin, and F. Jacob (1982). *EMBO J* **1**: 901.

758. J. K. Heath and M. J. Deller (1983). *J. Cell Physiol.* **115**: 225.

759. A. R. Rees, E. D. Adamson, and C. F. Graham (1979). *Nature* **281**: 309.

760. J. Heath, S. Bell, and A. R. Rees (1981). *J. Cell Biol.* **91**: 293.

761. C. M. Isacke and J. K. Heath (1982). *Biochem. J.* **208**: 235.

762. A. Rizzino, L. S. Orme, and J. E. de Larco (1983). *Exp. Cell Res.* **143**: 143.

763. J. Schindler, K. I. Matthaei, and M. I. Sherman (1981). *Proc. Natl. Acad. Sci. USA* **78**: 1077.

764. P. A. McCue, K. I. Matthaei, M. Taketo, and M. I. Sherman (1983). *Dev. Biol.* **96**: 416.

765. E. M. V. Jones-Villeneuve, M. A. Rudnicki, J. F. Harris, and M. W. McBurney (1983). *Mol. Cell Biol.* **3**: 2271.

766. M. K. S. Edwards, J. F. Harris, and M. W. McBurney (1983). *Mol. Cell Biol.* **3**: 2280.

767. C. Slack, R. H. M. Morgan, and M. L. Hooper (1978). *Exp. Cell Res.* **117**: 195.

768. M. L. Hooper (1982). In *Functional integration of cells in animal tissues* (J. D. Pitts and M. E. Finbow, eds.) (Cambridge University Press, Cambridge, U.K.), p. 195.

769. T. A. Smith and M. L. Hooper (1983). *Eur. J. Cell Biol. Suppl.* **1**: 38.

770. J. E. Seegmiller, F. M. Rosenbloom, and W. N. Kelly (1967). *Science* **155**: 1682.

771. M. L. Hooper and C. Slack (1977). *Dev. Biol.* **55**: 271.

772. J. R. Slack, R. H. M. Morgan, and M. L. Hooper (1977). *Exp. Cell Res.* **107**: 457.

773. M. J. Dewey, D. W. Martin, Jr., G. R. Martin, and B. Mintz (1977). *Proc. Natl. Acad. Sci. USA* **74**: 5564.

774. K. Illmensee, P. C. Hoppe, and C. M. Croce (1978). *Proc. Natl. Acad. Sci. USA* **75**: 1914.

775. E. F. Wagner and B. Mintz (1982). *Mol. Cell Biol.* **2**: 190.

776. A. J. J. Reuser and B. Mintz (1979). *Somat. Cell Genet.* **5**: 781.

777. M. J. Rosenstraus, R. F. Balint, and A. J. Levine (1980). *Somat. Cell Genet.* **6**: 555.

778. M. J. Rosenstraus, M. Hannis, and L. J. Kupatt (1982). *J. Cell Physiol.* **112**: 162.

779. M. J. Rosenstraus (1983). In ref. 711, p. 561.

780. S. Aizawa, L. A. Loeb, and G. M. Martin (1983). *J. Cell Physiol.* **115**: 9.

781. T. Boon and O. Kellerman (1977). *Proc. Natl. Acad. Sci. USA* **74**: 272.

782. M. Georlette and T. Boon (1981). *Eur. J. Cancer Clin. Onc.* **17**: 1083.

783. M. S. Featherstone and M. W. McBurney (1981). *Somat. Cell Genet.* **7**: 205.

784. C. MacDonald and G. Mosson (1983). *Cell Biol. Int. Rep.* **7**: 447.

785. B. W. Finch and B. Ephrussi (1967). *Proc. Natl. Acad. Sci. USA* **57**: 615.

786. M. W. McBurney, J. Craig, D. Stedman, and M. Featherstone (1981). *Exp. Cell Res.* **131**: 277.

787. F. J. Benham, M. V. Wiles, and P. N. Goodfellow (1983). *Mol. Cell Biol.* **3**: 2289.

788. M. W. McBurney and B. Strutt (1979). *Exp. Cell Res.* **124**: 171.

789. S. Linder, H. Brzeski, and N. R. Ringertz (1979). *Exp. Cell Res.* **120**: 1.

790. R. G. Oshima, J. McKerrow, and D. Cox (1981). *J. Cell Physiol.* **109**: 195.

791. T. Atsumi, Y. Shirayoshi, M. Takeichi, and T. S. Okada (1982). *Differentiation* **23**: 83.

792. J-P. Rousset, D. Bucchini, and J. Jami (1983). *Dev. Biol.* **96**: 331.

793. K. Illmensee and C. M. Croce (1979). *Proc. Natl. Acad. Sci. USA* **76**: 879.

794. D. Duboule, C. M. Croce, and K. Illmensee (1982). *EMBO J* **1**: 1595.

795. T. Watanabe, M. J. Dewey, and B. Mintz (1978). *Proc. Natl. Acad. Sci. USA* **75**: 5113.

796. F. Kelly and H. Condamine (1982). *Biochim. Biophys. Acta* **651**: 105.

797. A. J. Levine (1982). *Current Topics Microbiol. Immunol.* **101**: 1.

798. S. Segal, A. J. Levine, and G. Khoury (1979). *Nature* **280**: 335.

799. A. Linnenbach, K. Huebner, and C. M. Croce (1981). *Proc. Natl. Acad. Sci. USA* **78**: 6386.

800. L. Dandolo, M. Vasseur, C. Kress, J. Aghion, and D. Blangy (1980). *J. Cell Physiol.* **105**: 17.

801. F. K. Fujimura, P. E. Silbert, W. Eckhart, and E. Linney (1981). *J. Virol.* **39**: 306.

802. L. Dandolo, D. Blangy, and R. Kamen (1983). *J. Virol.* **47**: 55.

803. M. Vasseur, M. Katinka, and C. Marle (1983). In ref. 711, p. 259.

804. C. Stewart, H. Stuhlman, D. Jahner, and R. Jaenisch (1982). *Proc. Natl. Acad. Sci. USA* **79**: 4098.

805. J. W. Gautsch and M. C. Wilson (1983). *Nature* **301**: 32.

806. O. Niwa, Y. Yokota, H. Ishida, and T. Suguhara (1983). *Cell* **32**: 1105.

807. P. Herbomel, B. de Crombrugghe, and M. Yaniv (1983). In ref. 711, p. 285.

808. E. Linney, S. Donerly, B. Olinger, M. Bender, and F. K. Fujimura (1983). In ref. 711, p. 271.

809. J. Jami, C. Lasserre, D. Bucchini, R. Lovell-Badge, J. Thillet, M-G. Stinnakre, F. Kunst, and R. Pictet (1983). In ref. 711, p. 487.

810. K. Huebner, A. Linnenbach, P. K. Ghosh, A. ar-Rushdi, H. Romanczuk, N. Tsuchida, and C. M. Croce (1983). In ref. 711, p. 343.

811. K. Illmensee (1982). *J. Cell Physiol. Suppl.* **2**: 117.

812. J. McGrath and D. Solter (1983). *Science* **220**: 1300.

813. P. Newmark (1983). *Nature* **303**: 363.

814. E. Lacy, S. Roberts, E. P. Evans, M. D. Burtenshaw, and F. D. Costantini (1983). *Cell* **34**: 343.

815. R. D. Palmiter, R. L. Brinster, R. E. Hammer, M. E. Trumbauer, M. G. Rosenfeld, N. C. Birnberg, and R. M. Evans (1982). *Nature* **300**: 611.

816. G. S. McKnight, R. E. Hammer, E. A. Kuenzel, and R. L. Brinster (1983). *Cell* **34**: 335.

817. R. L. Brinster, K. A. Ritchie, R. E. Hammer, R. L. O'Brien, B. Arp, and U. Storb (1983). *Nature* **306**: 332.

INDEX

167